DESTINATION MARS

ROD PYLE

Foreword by ROBERT MANNING
Mars Science Laboratory Project Chief Engineer

DESTINATION MARS

New Explorations of the RED PLANET

59 John Glenn Drive
Amherst, New York 14228–2119

Published 2012 by Prometheus Books

Cover design by Nicole Sommer-Lecht

Inquiries should be addressed to
Prometheus Books
59 John Glenn Drive
Amherst, New York 14228–2119
VOICE: 716–691–0133
FAX: 716–691–0137
WWW.PROMETHEUSBOOKS.COM

16 15 14 13 12 5 4 3 2 1

Library of Congress Cataloging-in-Publication Data

Pyle, Rod.
Destination Mars : new explorations of the Red Planet / by Rod Pyle.
p. cm.
Includes bibliographical references and index.
ISBN 978-1-61614-589-7 (pbk. : alk. paper)
ISBN 978-1-61614-590-3 (ebook)
1. Mars (Planet)—Exploration. 2. Mars (Planet)—Surface. 3. Artificial satellites—Mars (Planet). 4. Space flight to Mars—Planning. I. Title.

TL799.M3P95 2012
629.43'543—dc23

2011050583

Printed in the United States of America on acid-free paper

To my parents, for understanding,
and to Sherry Clark, for knowing.

Love to you all.

CONTENTS

FOREWORD

The act of exploration is not what I thought it was.

I have a hard time reconciling my childhood memories of the birth of space exploration with the reality that I have experienced as a professional in the field of robotic space exploration.

I just barely remember the drama of John Glenn's heat shield in February 1962. I was too young to remember the play-by-play, but I followed the events closely a year or two later in grade school as my teacher read aloud a *National Geographic* story. John Glenn had just completed an American first: he had orbited Earth three times in his tiny Mercury spacecraft. As the last orbit approached, the nervous ground-control team calmly informed him that the light on the console showed a heat-shield malfunction, which probably meant that the shield would not stay in place when he reentered Earth's atmosphere after his third orbit. No worries though. They also professionally suggested that to ensure that the heat shield remained in place, he should not jettison the retro-rockets that where strapped around the heat shield in a three-arm hug. The retro-rocket straps should prevent the heat shield from slipping off during the extreme heating of entry. Easy to say, harder to hear when in orbit!

I remember thinking about those three little straps and the light on the console that said something was wrong. I could see the straps melting away and finally releasing the retro-rocket pack that was centered on the shield. How did they know that it would work? How could he trust their opinion? What if the retro-rocket pack slipped off sideways and took the barely attached heat shield with it? The controllers were very smart, I told myself. They must have used some advanced math to show that there was no concern. They must have confidently assessed the situation and known that John Glenn would make it home only if he did not jettison the retro-rocket pack too soon.

OK. Now I know better. Yes, they were very smart. These people were focused and fearless; brilliant people who gave up a life of invention, entrepreneurship, and certainly far better pay to do something that no one else did. I really cannot say that they were selfless. In fact, they were selfish in a particular way. They wanted to be *the* people doing this. Not someone else. THEM. Pushing the envelope. Calmly feigning confidence as they told Glenn that the retro-rockets would hold on to the heat shield. Terrified, they nonetheless felt that their guess was the best guess. The best guess from anywhere on Earth. And the best guess was *probably* right and was the way to success. They wanted to be the ones who were right. Being right was worth the low government pay. But the truth is that they did not know. They *could* not know. They were human, and humans know only so much.

So what do I know now? I know that space exploration is as exciting and as hard as anything humans have ever done. I think I sensed that in 1963 when I learned of this story and others that were playing out on black-and-white television screens across the country, including my family's TV. What I know now and what I have come to know for a long time is that space exploration is a deeply human endeavor.

But the people who envision doing science on another world, the people who invent these machines and instruments, the people in the back rooms with the white shirts and black ties are not "rocket scientists." They are simply people much like you.

They are optimistic, can-do, hopeful, bright, and sometimes quite lucky. But they are definitely human.

In the latter two-thirds of my career, I have been a Mars explorer. I have learned about both the amazing things people can do as well as our own limits as human beings. Perhaps that is what I did not know when I was a young newcomer to space exploration. I did not know about the moving boundary between what is possible and what is not. I did not know that every new idea, every new experience, every new mission was another layer that builds a foundation and pushes that boundary further and further aloft.

I have been very lucky to have witnessed and participated the in the drama of Mars exploration for the past twenty years. I have witnessed the veil of the known being parted with each new mission. Whether it is a scientific discovery of vast water deposits just under the Martian surface or a new engineering insight that tells us about better ways to land on Mars, these insights have built a remarkable era of discovery. Mars is not the mystery it once was, but it has evolved into a living place with dramatic vistas and secrets just below the surface. The missions and layers of discovery you will read about are real, made true by people who are driven by a deep curiosity and who are unafraid to go.

Perhaps we *are* like John Glenn and the explorers of my childhood after all.

Robert Manning
Mars Science Laboratory Project Chief Engineer
November 2011
Pasadena, CA

ACKNOWLEDGMENTS

There are many people to thank for their contributions to this book, and I hope that I have recalled you all.

First, I want to thank the excellent team at Prometheus Books. Linda Greenspan Regan, Steven L. Mitchell, Jade Zora Ballard, Ian Birnbaum, and Jennifer Tordy were all magnificently helpful and supportive. Meghan Quinn handled publicity with mastery. Catherine Roberts-Abel shepherded the book through its many versions, and Laura Shelley provided expert indexing (and, as it turned out, additional proofreading) services.

John Willig, of Literary Services Inc. and my agent, made the book a reality and was wonderfully and unendingly optimistic and supportive throughout. Alex Aghajanian, lifelong friend and attorney, lent his services as always.

The folks at Jet Propulsion Laboratory deserve a major tip of the hat. Rob Manning was supportive and encouraging, and carved out some of his very limited spare time to contribute both a chapter and the foreword for the book—and all this in the midst of readying the Mars Science Laboratory for launch. Guy Webster and Elena Mejia provided access to some of the top minds in Mars exploration today for interviews. And, of course, the people who labor countless hours behind the scenes to provide terabytes of data on the US Mars program online deserve recognition—it is, in my opinion, the finest data repository of its kind anywhere.

Robert Brooks, also of JPL, is a friend of many years and spent a number of hours with me, guiding me through the sometimes-Byzantine history of Mars exploration at NASA, as well as contributing to this book.

Loma Karklins at the Caltech archives was unstintingly helpful in finding somewhat obscure material from that institution's glorious past. Without her assistance, most of the interview material prior to Mars Pathfinder would be absent.

Chip Calhoun from the American Institute of Physics, Niels Bohr Library and Archives, also contributed to the archival efforts. He and the rest of the institute staff gave many hours of assistance retrieving material that is available nowhere else.

Many top researchers and planetary scientists gave me their limited and valuable time for interviews. In no particular order, they are:

Dr. Peter Smith of the University of Arizona
Dr. Steve Squyres of Cornell University
Dr. Joy Crisp of JPL
Dr. Richard Zurek of JPL
Dr. Chris McKay of NASA Ames Research Center
Dr. Laurence Soderblom of JPL
Dr. Robert Zubrin of the Mars Society and Pioneer Astronautics
Dr. Jeffrey Plaut of JPL
Dr. Bruce Murray of Caltech

And, posthumously:

Dr. Robert Leighton of Caltech
Dr. Norman Horowitz of Caltech

Gloria Lum provided expert grammar checking and creative input for the text, as she always has for my books, as well as unselfish support all around. Emil Petrinic gave the manuscript a thorough fact-checking, as did Bob Brooks, Dr. Jack Giuliano, and Robert Manning. True friends all. Jason Clark spent countless hours transcribing interviews late into the night.

Ken Kramer, friend of thirty-five years and a professionally trained psychotherapist, doubtless utilized some of his education in our many late-night chat sessions during the authoring process. Likewise Rodman Gregg, film producer, and Scott Forbes, entertainment professional. My son, Connor Pyle, gave up many evenings with his dad so that I could indulge myself in the magnificent mystery of writing about something I love. Leonard David, space journalist par excellence, lent support and the occasional answer to the unanswerable. Likewise Andy Chaikin, author of some of the best space-history books of all time. Jeanie and Joe Engle of NASA receive the same credit.

And to the folks who agreed to read galley copies of the book: Dr. Steven Dick, formerly NASA's chief historian; Roger Launius, senior curator at the Smithsonian Institution; Steven Fentress of the Strasenburgh Planetarium; Tony Cook of the Griffith Observatory; Leonard David, a premiere space journalist; and Piers Bizony, bestselling space author.

Book writing is a solitary yet collaborative experience, and without the advice, assistance, and support of these people, such efforts would not be possible. My heartfelt thanks to you all.

CHAPTER 1

THE FIRST MARTIAN

July 20, 1976: The Viking 1 orbiter instructed its lander to begin the separation sequence to start the long journey to the Martian surface. It was just after midnight at the Jet Propulsion Laboratory (JPL) in Pasadena, but as the probe was automated, no commands had been exchanged for some time. The onboard computer initiated a final round of systems checks. The explosives that joined the lander to the orbiter were armed . . .

Anxious flight controllers, largely powerless at this distance, could only watch the time-delayed data as the onboard computers made their own decisions. At 00:00 onboard computers fired the pyrotechnics, separating the Viking lander, which soon fired its own braking thrusters to begin the slow fall out of Martian orbit. In the dusky skies above, the orbiter from which it had recently separated continued on its mission. Below spread the ruddy expanse of Mars: dusty, cold, unexplored . . . and in about three and a half increasingly turbulent hours, *home*.

The Viking 1 lander, at ten feet wide by seven feet tall, was part of the largest and most expensive US unmanned mission to date. The orbiter, eight feet wide and ten tall, with a solar-panel span of thirty-two feet, shared the distinction. In a few weeks, Viking 2, a virtual twin, would arrive on Mars on an identical mission, but within a different landing zone on the opposite side of the planet.

The people who had sent Viking to this dangerous rendezvous waited out the landing confirmation signal in tense quiet. Only the most necessary words were spoken. There was an eighteen-minute delay between Earth and Mars at this distance; whatever happened to Viking now would be of its own doing. Many scientists on this program estimated a 50-50 chance of success, even with two landers. It was, in essence, a blind landing on a rocky, undulating landscape.

The Viking 1 lander was, for the first time in its short life, completely alone.

The tiny craft plummeted into the thin Martian atmosphere at 10,000 mph, still firing its braking thrusters. These rockets were models of simplicity. The fuel was a monopropellant and needed no ignition source and no other chemical mixed with it to explode into thrust. Further, instead of using complex pumps to feed the engine, the propellants were pressurized by stored helium gas. There was little to go wrong once they fired.

The lander was encased by a heat-resistant aeroshell, a dish-shaped structure that protected it from the heat of entry but also placed more demands upon its small digital brain. For as it plummeted through the upper reaches of the tenuous Martian atmosphere, Viking's computer was focused not just on a successful landing but also on conducting research in this wispy environment. Nothing is wasted in space exploration, and this early descent phase was no exception. As the computer labored to steer the craft, data began flowing in from sensors mounted on the aeroshell, providing data about charged particles surrounding the descending craft. Within the parade of arcane obsessions in the mind of the planetary scientist, understanding how the solar wind—high-energy particles streaming forth from the sun—interacts with the upper reaches of the Martian atmosphere is a thrill. The measurements now being recorded on the onboard tape drives should shed some light on this question. But Viking cared not; it simply stored the data for eventual delivery to Earth. Recording data was its raison d'être, and to this task it applied itself from its first moments.

At about 180 miles in altitude, another instrument switched

on: the mass spectrometer. This would measure the makeup of the upper atmosphere, analyzing the thin gasses present to provide a more detailed accounting of the "air" to augment the painstakingly gathered information already gleaned from Earth-bound telescopes. This first US spacecraft to enter another planet's atmosphere would accomplish multiple objectives, but primary among them was searching for one capable of supporting life as we understood it in 1976.

At about sixty miles high, this group of instruments switched off and another set became active. These performed an elegant analysis of the pressure, density, and temperature of the lower atmosphere by measuring the slowing of the craft. It was a bit like a waltz with a nonexistent partner, where one's success is measured via self-observation rather than direct feedback from the surroundings. But it was enough.

At about seventeen miles, the trajectory shifted: the aeroshell was sufficiently aerodynamic that it began to generate some lift, and Viking began to glide across the Martian sky. All this was by design; it was another way to scrub off excess velocity. Eventually, weight and drag took their toll and the craft began its steep descent once more.

The continual hiss of the rockets was joined by the roar of the thickening atmosphere, which, while thin, would soon be enough for the single parachute, set to deploy at nineteen thousand feet, to slow the machine sufficiently to land in one piece. This slowing to a sane rate of descent would be aided by more rocket engines. These were ingeniously designed as three clusters of eighteen tiny nozzles that would provide adequate braking propulsion without disturbing the surface upon which it alighted. All this, plus the fanatical sterilization of the spacecraft, was critical to preserving the sanctity of the ground below. For this was central to its primary mission—the search for life.

The onboard radar was scanning the ground, providing excellent data for range to the surface. What it was *not* providing was any idea of how rough that surface might be. The Viking team back on Earth had searched for the best landing place it could find with Mariner 6 and 7 photographic surveys, and later with results from Mariner 9, but it was barely better than a rough guess. At the Mariner 9 camera resolutions, the best images heretofore available, items smaller than the Rose Bowl were nearly invisible. Anything smaller than that had to be inferred from the analysis of surrounding terrain, and this was more alchemy than science, based on Earth-bound geological assumptions. Teams had agonized over these images for years. Then, data from the just-arrived Viking orbiter cameras resulted in more eleventh-hour angst about the landing area and a new site was selected at the last moment. Now all JPL controllers could do was aim the gun, close their eyes, and squeeze the trigger. In short, Viking was what lab folk later referred to as a BDL—a Big, Dumb Lander. Much of what happened from now on was based on luck. Viking could crash and mission control would be blissfully unaware until eighteen minutes after the fact, when the signal would simply vanish.

Soon the lander unhooked from the parachute, now relying only on its tiny landing rockets to control the final descent. At three hundred feet up, low-level radar kicked in to give a last set of readings. At sixty feet, the computer worked to cancel any horizontal motion and the lander settled into a strictly downward mode. It would now land directly below, no matter what. So said the simple instructions burnt into its primitive memory, saved in tiny magnetic cores that lived at the intersection of minute, hair-thin wires. While brutishly dumb by today's standards (your toaster probably holds more data), it was an elegant and almost bombproof method of storing data.

Slowly, Viking descended the final few feet. The rockets would not shut off until the lander made ground contact. But what lay below? The Viking lander had a scant 8.5 inches of ground clearance; any rock larger than that would likely end the mission. Falling in the weak gravity at a leisurely 6 mph, about the speed a person can walk, Viking 1 settled onto Chryse Planitia, Greek for "Golden Plain," a large and relatively flat expanse not far from the Tharsis volcanic region.

Touchdown. Silence returned to Mars. The Viking 1 lander was down, alive and well, after a 440,000,000-mile journey.

The date was July 20, 1976, the seventh anniversary of the landing of Apollo 11 on the moon. It was the first US soft landing on another planet (a moon is a satellite), and the first probe to function for more than a minute on another planetary body (an earlier Soviet probe had landed, but failed upon touchdown).[1] In fact, it would perform well beyond its builders' wildest expectations.

As the lander began surface operations, the Viking orbiter continued overhead, entering a new phase of its own science program. Armed with high-resolution cameras, it continued its observations while also acting as a relay station between the lander below and Earth, a blue star barely visible over the horizon.

Lander 1 went through a deliberate cycle of making sure that the descent engines and associated systems were shut down. It would not do to drip anything caustic or polluting onto the ground below. Hydrazine, the craft's volatile and corrosive fuel, would not be friendly to any microorganisms lurking about and would be a terrible way of saying hello. In fact, not so much as a microbe of Earth biota had been knowingly allowed to fester on Viking either; it had been baked, purged, and sterilized better than any surgeon's tool before launch. Nothing could be allowed to pollute the virgin Martian soil.[2] As the engines were "safed," the computer queried the navigation system, or inertial guidance unit. This simple system, while no longer needed for steering the craft, would help to supply altitude and directional information, so it was run for another five minutes. This information was critical to aiming the radio dish toward Earth, so the more accurate the data, the better.

At the same time, the first postcard to home was being assembled. The Viking landers used a new type of imaging camera. Previous space probes had used state-of-the-art TV cameras, but at the time, the images were not up to what the designers had yearned for. For Viking, the camera stared upward into a mirror that swung vertically, "nodding" up and down. Between each nod the mirror would rotate a small amount. In this way, a series of strips were assembled over time, and these resulted in what was, for the day, a very high-resolution image. Two of these ingenious devices were mounted on each lander, allowing three-dimensional imaging, and the first job of the day was to send an image home.

But this first snapshot of another planet was not to be a splendid panoramic of the landing area; rather, it was a somewhat mundane image of the nearest footpad. This would accomplish multiple goals instantaneously: the safety (or lack thereof) of the landing site would be demonstrated by the placement of the footpad. The amount of sinking into the sandy soil (properly called *regolith*, as the word *soil* implies life within) would be shown, and this, along with other measurements such as the amount of slowing at contact and the designed-in collapsing of the lander's legs upon touchdown would supply information about the compactibility of the ground. Remember, nothing is wasted in space exploration.

Back on Earth, strips of the first picture from Mars began to

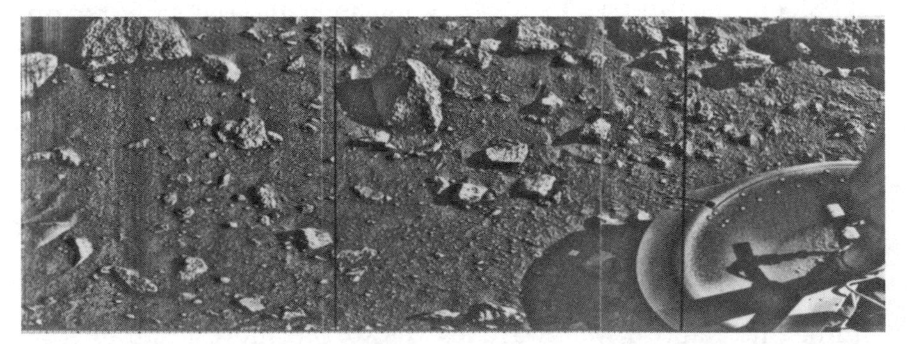

Figure 1.1. FOOTPAD: This very first image from the surface of Mars shows one of Viking 1's footpads planted securely on the surface. It was important to mission managers to know how well placed the craft was. *Courtesy of NASA/JPL.*

come in. It was innocuous enough: a shot of footpad 3. If the probe had failed then and there, a lot of folks would have been very upset to have nothing more to show for the billion-dollar effort. But this shot was needed to ensure that the craft was stable. Cheers rang out at JPL and Caltech as the proof of a successful landing were made visible. But from Mars, the lander could not hear, nor would it have cared. It merely carried on in its eighteen kilobytes of programmed duties with dogged and ruthless determination.

Next on the lander's to-do list were the pyrotechnic events, known to most of us as explosions. In spaceflight, whether manned or unmanned, small explosives had long had a leading role. Then as now, they were used to separate the stages of rockets as they ascended away from Earth. They released spacecraft once in orbit. They opened and closed valves. And, in Viking's case, they were critical to beginning Mars-based activities. These are, by their nature, one-shot operations—as in, they work or they don't. Their duties included releasing safeties for the life-science experiments and opening the meteorology boom—an arm with instruments to measure wind speed, temperature, and the like. These performed without a hitch.

Now a second photo was taken, and this was the money shot: the first picture of the horizon of Chryse Planitia. As the lander went about its business, breath was again held in mission control. What would we see? What did the surface of Mars look like at ground level? Remember that these were the days of rotary telephones, bias-ply tires, and such state-of-the-art things as *The Eagles: Greatest Hits* via vinyl records. An image from the surface of Mars was heady stuff. And with the Viking orbiter disappearing over the horizon in about twelve minutes, and with it, the best link to home, this had to be done *now*.

Once again, Viking 1 did not disappoint. The first image, black and white but glorious nonetheless, slowly assembled, again, a strip at a time. The tension broke slightly as the first strip came in, but like a good mystery novel, Mars was only revealed a small bit at a time. The results were well worth the wait. After years of preparation, a billion dollars, and a journey of many times the 119 million miles then separating Earth from Mars, the first landscape was in. The data was still coming back long after the orbiter was out of touch, given the long transmission travel time across the vast darkness, and the lander went into a base-operation mode while out of communication.

But the picture . . . oh, that second picture. It lacked color and was obscured on the bottom by various parts of the spacecraft. But there it was, in all its monochromatic glory: the horizon of Mars. Low, arid hills were off in the distance, and between the lander and those hills was an expanse of sharp, jagged rocks. Hundreds of them. And off to the right, dominating the horizon there, was the bright glow of the sun, unseen and above the frame. It was a dry, cloudless spectacle. For someone seeking the serenity of an English tea garden, or the Mars of Percival Lowell, it would not do. But for any human pining for a glimpse of another world, a world we could relate to, another planet to which we might one day travel, it was nirvana.

Viking 1 was, however, oblivious to such human emotion. The outbursts and cheers from Earth remained unheard. It had a primary mission of just sixty days on the surface, with an extended mission target of 120. At that point, Mars would pass behind the sun and communication would be lost for weeks. And while controllers on Earth planned to "safe" the lander during this time, their confidence in reawakening the machine after this period was limited. But true to what would become JPL's legacy of performing near miracles with distant machines, the first lander operated successfully for well over *six years*. And the tale of its ultimate demise is not one of equipment failure, but of human error.

With Viking's successful landing, there was now time—well over two months in the primary mission alone—to perform the tasks it was designed to do. The instructions came up from Earth in carefully coded batches, to be processed and executed in sequence. With mechanical exactitude, Viking 1 began its primary labors—taking color images of the surrounding surface, digging scoops of soil and dumping them carefully into small funnels that led to an onboard laboratory, and fulfilling its primary objective: the search for life on Mars.

Just under two months later, on September 3, 1976, the Viking 2 lander settled gently onto Utopia Planitia, 4,200 miles away to the northeast. Humanity now had two outposts on Mars, and the exploration of the red planet began in earnest. Overhead, the Viking orbiters continued to chip away at their intense workload, snapping pictures and sending reams of data earthward. What they imaged and reported would change our understanding of Mars overnight: the Martian Renaissance had begun.

CHAPTER 2

MARS 101

Exploring Mars is a bit like doing brain surgery through a mile-long soda straw. At an average distance of fifty million miles from Earth, with a one-way radio message time of twelve to twenty minutes, roving the dry, treacherous surface requires the utmost in planning and careful execution. One false move can end a mission in seconds, and there are rarely many options to correct a mistake. That is why the people who dare seek the truth about Mars are so remarkable. This is the story of human striving, from early times through tomorrow, to discover what makes Mars tick.

Orbiting in the dark cold of space at an average of about 140 million miles from the sun, or about half again as far as Earth, Mars is the fourth planet in the solar system. It is also the last stop for the rocky, or terrestrial, worlds before the gas giants Jupiter and Saturn and the icy balls of Uranus and Neptune. It is separated from these giant worlds by the asteroid belt, a planet which failed to form from the large disk of material that still orbits the sun beyond Mars.

The air on Mars is thin and cold; the highest temperatures hover at about 60°F, and can plummet to −180°F at the poles. Its day lasts about 24 hours, 37 minutes, and its axial tilt matches

Earth's at 24 degrees. Its year lasts 686 Earth days. Mars is about half the diameter of Earth, less than a quarter its size, and has only 11 percent of Earth's mass and 38 percent of its gravity. Despite this, it has almost the same dry surface area, due to a lack of seas and oceans; in fact, bodies of liquid water do not exist on its surface. It is a bone-dry, ultracold place with only about 1 percent of Earth's atmosphere.

Why then, one might ponder, are we so fascinated with this seemingly inconsequential world? Because Mars is a planet of dreams. Always, it has inspired feelings in humanity as no other planet in our solar system. Early on, it was the reddish hue, resembling the color of aging blood, that attracted the naked eye. Later, in wavering telescopic images transmitted across the tens of millions of miles from its surface, it was the odd markings and, still later, imagined lines crossing its surface that inspired. One could imagine life there. One could imagine . . . empires.

And why not? The planet is not that far from Earth, and must then be not so unlike our own world, or so the thinking went. If Venus, one step closer to the molten sun, might be covered with tropical oceans and riotous growths of green, steaming jungles under its impenetrable cloud cover, why couldn't Mars, still within the so-called Goldilocks Zone, harbor an older, wiser, more advanced civilization?[1]

But of course, those dreams vanished along the way. Venus turned out to be an unbelievably hot hellhole with over nine hundred pounds per square inch of pressure crushing its deadly surface. But the real Mars, as we know it today, is not so much less interesting than the one of previous generations. The empires of Edgar Rice Burroughs's green men and eight-legged Thoats (dinosaurlike steeds) might be gone, but in its place is an old, highly weathered, and geologically fascinating place with signs of vast and ancient floods of liquid water, and more recent indications of smaller flows.

The planet seems in many ways to be an older version of our

own. But it is an Earth with planetary evolution gone awry. Once rich with oceans and cloud cover, it is too small to any longer harbor liquid water on its surface for more than a few moments. Its thin atmosphere is almost entirely carbon dioxide, with bits of nitrogen, argon, and oxygen existing in wisps. Most of the life-sustaining oxygen that we prize so highly has been long spent, slowly turning the iron in the soil a ruddy, oxidized red. And that thin atmosphere has also allowed for billions of rocks, most of which would burn up or explode in Earth's denser atmosphere, to slam into Mars's surface with impunity. Many millions of these were large enough to leave wounds on the planet, and some created vast new surface features.

Another seeming indicator of a dead world is Mars's lack of a meaningful magnetic field. This is probably due to a largely inert core, or a cooling one. Whatever the case, there is not the same molten, metallic dynamo that creates Earth's robust lines of magnetic force, and what magnetism Mars does possess is lumpy and erratic. The gravitational field is also unlike Earth's, with mascons (mass concentrations), not unlike those within Earth's moon.[2] Whatever the case, Mars has, proportionally, a much thicker crust and a smaller, less active molten core than our planet.

Still, Mars has possessed life, though not the kind we seem to wish upon it. It was *geologically* alive, with vast lava flows and wind and water erosion working their healing magic upon its tortured and pockmarked surface. While much of the southern half of the planet still bears the scars of bombardment, the northern half is largely covered with younger lava flows that filled in the offending craters. And the source of much of this once-molten rock can be clearly seen with even low-resolution cameras from the myriad probes that have flown past the planet, in the form of the Tharsis Bulge and its huge volcanoes. This region is so remarkably swollen that it noticeably deforms the otherwise spherical planet. And it is home to some of the most impressive mountainous real estate in the solar system.

What follows is a brief primer of Martian geography. Entire sets of texts are available on the subject; what is presented here is merely the briefest of samplings. The intention is to present a general idea of the important regions of the planet in both historical and scientific terms.

First among the huge volcanoes identified within the Tharsis region is Olympus Mons, the largest known mountain anywhere, which flanks the bulge. Three times the height of Mount Everest, the now-extinct volcano soars fourteen miles into the thin Martian air. It is a shield volcano, resembling those that comprise the islands of Hawaii, with a base width of almost four hundred miles. The area it rests upon is roughly the size of Arizona. It is also the youngest of the major volcanic structures on Mars.

Directly atop the bulge and spanning its crest diagonally are three older volcanoes, all shield volcanoes, Arsia Mons, Pavonis Mons, and Ascraeus Mons. While subordinate in size to their larger sibling, these lava factories contributed greatly to the basaltic flows that inundated much of the surrounding area. Overall, the Tharsis region is the size of a terrestrial continent.

The formation of this region was not without its side effects, and a gigantic wound in the planet can be found nearby, stretching east from the Tharsis area and continuing along the Martian equator for about a quarter of the planet's circumference. This gigantic gash in Mars's hide is called Valles Marineris. In keeping with the Texas-style "bigger is better" nature of Martian topography, it is the largest valley in the solar system. Our own Grand Canyon would be scarcely noticeable alongside it. Almost 2,500 miles long, it was formed when the Tharsis region rose out of the planet, and the nearby crust could not take the stresses of this enormous violation. So it cracked and slumped, resulting in the huge channel. It averages 125 miles in width and is as much as 4.5 miles deep. It is outclassed only by the underwater Mid-Atlantic Ridge on Earth.

To the north rests Alba Mons, also known as Alba Patera, the oddest of the volcanoes and unlike anything else on Mars (or Earth, for that matter). It has the gentlest slope of any Martian volcano, with an inclination of just one-half a degree, or about a tenth of that of Olympus Mons. Its volcanic outflow forms an ellipse almost 2,000 miles across and 1,200 miles north to south, making it (here, again, the Texas-style attributes) among the largest known magma generators in the solar system. There are many theories seeking to explain its productivity, including the existence of highly fluid magma that flowed freely and fluidly out of the caldera and across the surface of Mars. On Earth, the total outflow of the volcano would have covered most of twelve states if centered in Colorado. It's a big beast.

Other volcanic regions include an area thousands of miles west of Tharsis called Elysium. Here we find three main volcanoes: Elysium Mons, Hecates Tholus, and Albor Tholis. Finally, down toward the equator is the region of Syrtis Major Planum with its own volcano of the same name. About 750 miles wide, but only one mile in elevation, this vast, low-lying monster produced lavas that seem to be different than that from the Tharsis volcanoes; it is more complex and differentiated in geological terms and is thought to have formed in the vast three-mile-deep magma chamber below when heavier elements settled out, leaving the lighter lavas to spew forth.

In any case, by the time Mariner 9 had begun sending back images of the Tharsis volcanic complex, any thoughts that Mars had been dull or uninteresting in its youth were banished. While the planet may have slowed down in its old age, in earlier eras it was a geologically active toddler, with regular volcanic tantrums to match.

Pulling farther back, we can see that almost half the northern hemisphere is covered by the Borealis Basin. Its origins are uncertain, but it was likely the result of a huge, planet-shifting impact. What is apparent is that there are far fewer craters in this area, and the Tharsis Bulge was formed subsequent to the events that spawned Borealis. If it is an impact feature, it would again be a record setter as the largest in the solar system. And the object that impacted Mars would have been about the size of Pluto, and probably arrived during the Late Heavy Bombardment period of about four billion years ago.

In the southern hemisphere can be found Hellas Planitia, another huge impact basin, about 1,430 miles wide with a depth of about 30,000 feet. It is so deep that the atmospheric pressure at the bottom is about 90 percent more than at the surface, enough to allow liquid water to exist for brief periods. While much smaller than the planet-girdling Borealis Basin, it is the largest obvious impact feature clearly visible on Mars. Like its northern cousin, it is thought to be about four billion years old.

Drawing a line from Hellas Planitia through Mars's interior to the other side of the planet, we return to Alba Patera, home of the Tharsis volcanoes. It is hypothesized that the impact at Hellas was sufficient to cause at least part of the formations on the antipodal, or opposite, side of the planet, as the seismic shock rattled through, slamming into the far side.

Much of the southern hemisphere, ranging a bit into the north, is extensively cratered, another result of the Late Heavy Bombardment, when copious amounts of interplanetary junk smashed into the rocky, or terrestrial, planets.[3] Overall, the southern regions sit much higher than the northern hemisphere, and the crust of the planet is over twice as thick in the south.

The geological record of Mars can be summarized in three eras:

- The Noachian Period, 4.5 to 3.5 billion years ago, was when the oldest parts of the planet that remain were formed. These regions are covered with extensive, overlapping craters and show more of them than other areas due to their advanced age. The Tharsis Bulge formed during this period in Mars's early, and violent, history.

- The Hesperian Period, 3.5 to about 3 billion years ago, when basaltic magma flowed out from the planet's interior and formed the filled basins we see today.
- And the Amazonian Period, about 3 billion years ago through today. Olympus Mons and its associated lava flows were formed during this period, and less cratering is evident due to their younger age. This topography can be quite varied.

Maps of Mars list two general sets of features. The first are those differentiated by apparent brightness, called *albedo* features. Albedo is the amount of sunlight reflected back from another world. On maps, these have Latin names. One such example is the enormous Sinus Meridiani, or Meridian Bay, one of the few major features visible through a telescope from Earth. It is noteworthy that the darker of these features were originally thought to be seas or other bodies of water, and were named accordingly. Hence, some of the major features in this group are Mare Erythraeum (Erythraean Sea), Mare Sirenum (Sea of Sirens), and Aurorae Sinus (Bay of the Dawn). The largest dark feature seen from Earth is Syrtis Major Planum, a classical Latin name for a region near present-day Libya. Areas thought at the time to be dry land include Arabia Terra (Land of Arabia) and Amazonis Planatia (Amazonian Plain). The north polar cap is Planum Boreum (Nothern Plain), and the southern cap is Planum Australe (Southern Plain).

Now it is time to discuss dirt. Earth has dirt, also called *earth*. But on Mars, and other solid bodies like the moon, one cannot properly refer to the soil as dirt. The proper term is *regolith*, from the Greek *rhegos* or blanket, and *lith* or rock. It denotes a layer of loose material over bedrock, essentially ground-up rocks. Spacecraft have studied or observed the regolith of our own moon, Titan (moon of Saturn), Venus, and Mars. In this book, however, we will generally refer to regolith as soil for the sake of expediency.

Martian soil is highly alkaline, and apparently filled with perchlorate. It is highly toxic stuff, at least so far as lower forms of life are concerned. But the areas of Mars sampled as yet are small, and orbital data inconclusive, so the true nature of planetwide soil is yet to be decisively determined.

Most of the planet is also overlain by a thin layer of oxidized dust, which gives Mars its red color. This dust is very finely grained and, when winds whip up, can stay suspended in the atmosphere for weeks or even months.

The planet has two icy polar caps, some of the first features to be observed through early telescopes. With a tilt of about 25 degrees, these polar areas experience growth in the local winter and shrinkage in the local summer. This led many eighteenth-century astronomers, most notably Percival Lowell, to conclude that poles were water ice and melted in the summers, sending life-giving water cascading to the equatorial regions of the planet and contributing to the "wave of darkening" that seemed to occur every Martian year, thought then to be plant life gone wild as the moisture from the poles nourished it. We now know that the poles are a mix of a thin layer carbon dioxide ice, or dry ice, and water ice. The CO_2 freezes groundward in the local summer and then is re-released into Martian skies in the summer. This seasonal exchange can account for over a quarter of the atmosphere freezing out, then being released back into the air. The water ice below also melts in the summer; evidence of flowing meltwater, once thought to be very unlikely, is seen in orbital images of the planet.

In fact, the observation of subterranean and frozen water is becoming relatively common on Mars. At one time thought to be dry, arid, and dead, Mars is turning out to be quite active in terms of weather and erosional processes. Wind is of course the primary agent of erosion today, but water is slowly, stubbornly revealing itself more and more.[4] The southern ice cap alone, the smaller of the two, is thought to have enough water tied up in its frozen reservoir to cover the entire planet to a depth of over thirty feet—but not for long. The atmosphere is so thin, at less than 1 percent of Earth's, that liquid water quickly boils away in the tenuous air. But scratch the surface, literally, and there is plenty of water ice all over the planet.

That water was far more evident in the past than it is today, slumbering beneath the dust, dirt, and sand of the present Mars. Huge features called *outflow channels*, about twenty-five of them, litter the surface of Mars, indicating a truly massive and disastrous outpouring of water at some time within the last few million years—and smaller ones are still being formed to this day, via liquid water. There are also areas strongly resembling river deltas, alluvial fans, and channels that speak to a once-watery Mars.

So what's the big deal about water? Simply this: life as we understand it needs water to exist. Increasingly, extreme forms of life on Earth are discovered that can survive on trace amounts and in hideously difficult environments, but all forms need some water. So when we find water in any form on Mars it is exciting, for it allows us to think, once again, of Martians . . . microscopic though they might be.

And why are microbes on Mars so important? It's not like they will be the next explosive market for Big Macs® or directly impacting our daily lives if discovered. But the discovery of some form of life on Mars would be a game changer in other ways. Philosophy, religion, and of course science would all be rocked by such a discovery. The existence of life on Mars would, for some, seem to diminish our special place in the universe as humans. Of course, if this life were something more akin to streptococci than your next door neighbor, it's hard to get too intimidated. But for some, any discovery of life elsewhere would be a threat.

For others, it would be a delight. The idea that life, that poorly understood miracle of amino acids and organic compounds, could have sprung up independently on another world is a big one. Some scientists have hypothesized that life may not have even begun here on Earth, that is may have started first on Mars and then hitchhiked a ride to Earth via a meteoric fragment. This idea has gained credibility in the last few decades, when meteors found on Earth (in Antarctica) were definitively traced back to Martian origin. And, since Mars calmed down from its evolutionary throes much sooner than Earth, this idea makes some sense. Life could have slowly evolved there, then come to Earth and survived once our own planet was less volatile. It would certainly give the organisms more time to mature in evolutionary terms.[5]

Of course, there are others who consider this to be hogwash, and still others who consider these ideas blasphemous. The former group will likely come around given sufficient evidence. The latter will never be convinced, regardless of the science; the idea is simply too threatening.

The important thing is this: until we go to Mars, sift through the soils, test its properties, and search cracks, crevices, caverns, glaciers, the polar caps, the equatorial regions, and everything in between for microbial (or other) life, we will probably not know the answer. And it may well take the hand of humanity present on the surface of that cold, ruddy world to make this work—robots are simply too limited and too inflexible.

It's almost time for us to visit the red planet. But first, let's look back a few millennia, to a time when Mars was not a destination, not yet even a plane, but a harbinger of death and destruction in the night sky.

CHAPTER 3

IN THE BEGINNING

A SHINING RED EYE

Of course, long before NASA's Vikings invaded Mars, the planet had been prominent in the mind of humanity. Viking was simply a logical outgrowth of our curiosity; a by-product of the heyday of space exploration in the 1960s and early 1970s. This was a time of great political rivalries when the United States and the Soviet Union sought to show the world which political system was superior in both technological and economic achievement. It was, true to the time, a flexing of national muscles and a brute-force approach to space exploration—a war without bullets or bombs. America won the moon first and, despite some close misses by the Soviets, Mars as well.

But the interest in Mars goes as far back as its visage in the night sky. Every major culture worshipped it until the Age of Reason; most commonly as a baleful godly eye. The late Babylonians addressed Mars as Nergal, representing fire, destruction, and war; in their system, Mars won out over the god of the sun, who was an earlier incarnation of Nergal, god of the underworld.

The Hindus referred to Mars as the deity Mangala, born of the sweat of Shiva (another cosmic troublemaker), meaning "auspicious," "a burning coal," and "the fair one." In Sanskrit, the name is Angaraka, a celibate (and probably frustrated) god of war. Egypt first called the red planet Horus Am Akhet (Horus on the horizon), then later Her Deshur (Horus the red). The city of Cairo is named after Mars, from Al Qahira, an ancient Arabic name for the planet.

Figure 3.1. NERGAL: The Babylonian god of war and pestilence is seen here as a man-lion. At one time a god of the sun, Nergal later became Babylon's representative of Mars. *Courtesy of iStockphoto.*

The ancient Chinese and Koreans saw Mars as a portent of bane, grief, war, and murder. It was named "fire star." This followed the Chinese mythology of the Five Elements: wood, earth, metal, water, and of course fire. One can guess where Mars fell in the lineup.

Plato's Greece called Mars Ares, the son of Zeus and Hera (of course he would be a warrior with these two for parents). Two of Ares's three children were Phobos (fear) and Deimos (terror), after whom Mars's two moons are named. Later Greece referred to the planet as Pyroeis, meaning "fiery." In either identification, it's difficult to get past the idea that Mars portended a bad day.

When Rome came onto the scene, it adopted the gods of Greece under new names. Ares became Mars, born of the goddess Rhea-Silvia, and producing two sons, Romulus and Remus. This led to the founding of Rome, so perhaps their warlike ways were preordained, as Mars was the most widely worshipped of their gods.

By the Middle Ages, Europeans characterized Mars's influence over humanity in more detail:

> Mars rules catastrophe and war, it is master of the daylight hours of Tuesday and the hours of darkness on Friday, its element is the fire, its metal is iron, its gems jasper and hematite. Its qualities are warm and dry, it rules the color red, the liver, the blood vessels, the kidneys and the gall-

> bladder as well as the left ear. Being of the choleric temper it especially rules males between the ages of 42 and 57.[1]

Interestingly, while they may have missed the boat regarding gallbladders and "daylight hours on Tuesday," NASA rovers have found hematite and, of course, oxidized iron in abundance on the Martian surface.

In any ancient culture, Mars was one of a handful of planets visible to the naked eye, and the only one of marked color, so the planet demanded attention. Its distance to Earth varies with two-year cycles, and its brightness waxes and wanes correspondingly to the intersection of these orbits. At the far end of this ever-changing distance it is over 250 million miles from our planet and a dim red star, distinguished only by its very un-starlike motions in the night sky (all visible planets move at rates different from the starry backdrop). At its closest, about 35–62 million miles from our world (depending on the year), it is the third-brightest object in the evening sky after the moon and Venus, Earth's other close planetary neighbor. Add to this that Mars has the most elliptical orbit of any planet (only Pluto's is more elongated, but that small body was recently demoted from the roster of planets in our solar system).

To further stand out from its astral competition, Mars did the remarkable: it moved backward from time to time. Called *retrograde motion*, when the Earth (which orbits inside the ellipse traveled by Mars) closes on Mars and passes it, Mars appears to go from a forward motion to a backward one ("retrograde") when viewed against the backdrop of stars.[2] Ptolemy came up with an explanation for this back in the Mars-worshipping days, but it assumed the Earth as the center of the universe and was mechanically flawed. Copernicus, as refined by Kepler, redesigned this mechanical explanation with the sun at its proper place at the center of the solar system, between 1510 and 1514 (the ancient Greek Aristarchus of Samos had posited this in the third century

BCE, but it was traded off for the Earth-centric model). Kepler also spent an additional eight years working out the elliptical nature of Mars orbit, with the sun at one focus-point of the ellipse.

In 1659, Christian Huygens, working in Holland, observed and drew Syrtis Major on a sketch of Mars. It was the first such recorded observation. Seven years later, in 1666, Giovanni Cassini measured the rotational period of Mars (i.e., its day) at 24 hours, 40 minutes (he was off by less than three minutes). Then, between 1777 and 1783, English astronomer William Herschel noted the axial tilt of Mars and deduced that it should have seasons not so different perhaps than Earth's.

No matter how rational the great thinkers were regarding Mars, a planet (especially the color of blood) which occasionally traveled backward was sure to gain notoriety. Remember that in a time when electric lights were unknown and the entire world fell under a carpet of darkness except for the flickering flames in individual dwellings, a red star above stood out much more markedly than it does in modern times.

But for our purposes it is the age of modern science, beginning in earnest in the 1800s, that is important. While men like Kepler toiled to develop concrete notions about the true nature of nearby

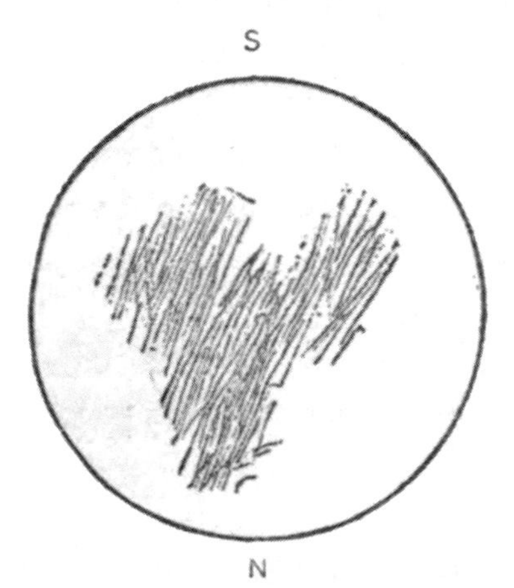

Figure 3.2. SYRTIS MAJOR: Though little more than a sketch, this early attempt at a map of Mars by Huygens displays the region of Syrtis Major, one of the few major features visible through an early telescope. *Courtesy of NASA.*

space and the planets (including his laws of planetary motion), the pace accelerated dramatically with the advent of the large optical telescope.

What was known about Mars at that time can be summarized thus:

- Fourth planet from the sun
- Smaller than Earth—about half of our planet's diameter
- Larger orbit than Earth's
- Thought to have no moons
- Thought to have oceans and continents
- Perhaps a thin, yet Earth-like, atmosphere

And, in a colossal misinterpretation of telescopic observations, one English astronomer stated in 1860, "There is no portion of the planet Mars that cannot be reached by ship."[3]

Mapping Mars came into vogue in the late 1800s as well, with the advent of improved optics, larger telescopes, and better tracking mechanisms. Astronomers like Camille Flammarion, an influential French scientist and spiritualist; Asaph Hall, an American in government employ; Giovanni Schiaparelli, who assiduously mapped Mars's surface through his telescope; and Percival Lowell, an American amateur astronomer, began to understand Mars as a planet. It was, however, a somewhat flawed vision.

Flammarion (1842–1926) was perhaps the more interesting of the lot, as he was not just an astronomer with his own observatory (as Lowell would become) but also an avid spiritualist who believed, among other things, that Martians were trying to communicate with Earth. He was an avid student of the occult sciences, as he called them, and had spent time as a young man studying in a Paris seminary. Flammarion apparently believed in reincarnation and was very spiritual. Yet, when it came to Mars, he at least made the attempt to remain somewhat objective, considering the era.

In 1873, he wrote:

> In Mars there is neither an Atlantic nor a Pacific, and the journey round it might be made dryshod. Its seas are [M]editerraneans, with gulfs of various shapes, extending hither and thither in great numbers into the terra firma, after the manner of our Red Sea. The second character, which also would make Mars recognizable at a distance, is that the seas lie in the southern hemisphere mostly, occupying but little space in the northern, and that these northern and southern seas are joined together by a thread of water. On the entire surface of Mars there are three such threads of water extending from the south to the north, but, as they are so wide apart, it is but rarely that more than one of them can be seen at a time. The seas and the straits which connect them constitute a very distinctive character of Mars, and they are generally perceived whenever the telescope is directed upon that planet.[4]

Then, in the same treatise, Flammarion teased his readers with rationality, only to dash it again:

> We speak of plants on Mars, of the snows at its poles, of its seas, atmosphere, and clouds, as though we had seen them. Are we justified in tracing all these analogies? In fact, we see only blotches of red, green, and white, upon the little disk of the planet; but, is the red, terra firma; the green, water; or the white, snow? Yes, we are now justified in saying that they are. For two centuries astronomers were in error with regard to spots on the moon, which were taken for seas, whereas they are motionless deserts, desolate regions where no breeze ever stirs. But it is otherwise as regards the spots on Mars.

Perhaps not surprisingly, he also wrote some early works that could be classified as science fiction (alien encounters, no less). It should be mentioned that in his spiritual writing he at least made an attempt to bring the scientific method to his efforts.

Asaph Hall (1829–1907) was a self-taught astronomer who nonetheless managed to wrangle a position at Harvard (he attended two colleges but never graduated from either). Despite his academic shortcomings, he managed to become at various times a professor of mathematics and the president of the American Association for the Advancement of Science. During his career, he authored over five hundred scientific papers about his astronomical observations, primarily about double stars and the planets. Working at the US Naval Observatory, Hall created maps of Mars and was the first to note its two moons, in 1877. The names Phobos and Deimos were suggested by a scholar at Eton, in Britain, as a fitting nod to Homer's *Iliad*.

Giovanni Schiaparelli (1835–1910) was far less eccentric than Flammarion and a more prolific mapmaker than Hall. He spent time observing on telescopes in locations as varied as Berlin and Russia, eventually returning to Italy. Besides his mapping of Mars, he is known for his discovery of the relationship between yearly meteor showers on Earth and the comets from which they originate. In the same year that Hall discovered the twin moons of Mars, Schiaparelli made some of his first detailed maps using his nine-inch refracting telescope in Brera, Italy. The lines he observed (or thought he did), and the derived markings on his maps, he termed "*canali*," or in Italian, channels. He later claimed that he did not intend this to suggest intelligent engineering: "[T]hese names may be regarded as a mere artifice. . . . After all, we speak in a similar way of the seas of the Moon, knowing very well that they do not consist of liquid masses."[5] Elsewhere, he continued: "[W]e are inclined to believe them to be produced by an evolution of the planet, just as on the Earth we have the English Channel and the Channel of Mozambique."[6]

However, Schiaparelli himself confused the issue, perhaps cautiously bowing to popular enthusiasm: "Their singular aspect, and their being drawn with absolute geometrical precision, as if they were the work of rule or compass, has led some to see in them the work of intelligent beings. . . . I am very careful not to combat this supposition, which includes nothing impossible."[7]

His maps were of sufficient quality that they were used well into the twentieth century, almost to the dawn of the space age. But the canali were the overriding feature to most. In the popular mind, and in at least one eccentric American astronomer's, the word *canali* had a specific connotation, and intelligence was indelibly ascribed to his observed features regardless of his ambivalence to the idea. This notion quickly caught fire in the public mind.

It should be noted that there was ample opposition to this idea in scientific circles. In 1894, American astronomer William Wallace Campbell (later to become the president of the University of

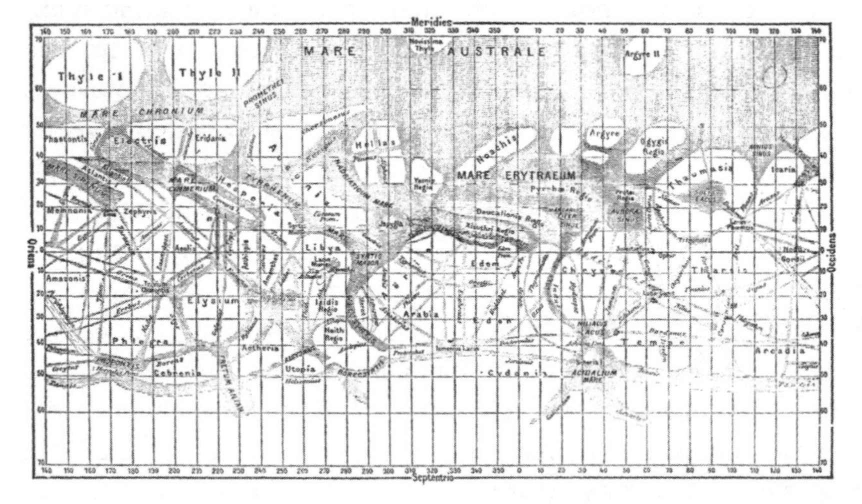

Figure 3.3. MARS, 1880s: Schiaparelli was one of the first to create highly detailed maps of Mars with classical names. These maps were used well into the twentieth century. For those who have looked at Mars through a telescope, it is clear that these early mapmakers needed great eyes as well as a fertile imagination. *Courtesy of NASA.*

California) performed spectroscopic analyses of the Martian atmosphere and noted little, if any, water or oxygen present. This should have crushed the dreams of many who envisioned advanced civilizations and vast oceans on Mars, or even water-consuming plant life, but this is not the way of the fertile and wishful mind. The idea of sentient Martians persisted, and grew.

The aristocratic American amateur astronomer Percival Lowell (1855–1916) was intrigued by a book Flammarion wrote about Mars (*La planète Mars*, 1892), and was further awed by Schiaparelli's work. In particular, Lowell seems to have fallen prey (as so many did) to the Italian astronomer's unfortunate choice of the word *canali* to describe the lines he thought he saw on the planet, thinking (or hoping) that this implied intelligent, intentional design. In short, Lowell and others interpreted this term, either consciously or not, as denoting artificial canals, built by intelligent Martians to save their water-starved world. In fact, Lowell went so far as to suggest a utopian, united Martian global government, for how else could they achieve these world-girdling civil-engineering projects?

Lowell was perhaps the most fascinating of the "intelligent Martian" club. Born of a wealthy Boston family, his passions were few but remarkably powerful throughout his life. After his formal education, which resulted in graduation from Harvard in mathematics, he spent many years in Asia, specifically in Japan, about which he wrote influentially and with characteristic intensity. This literary training would serve him well when he later sought to popularize his theories about Mars, which in their own way did more to popularize planetary science (no matter how ultimately misguided his own ideas may have been) with his populist, if fantastic, theories.

In 1894, with a few years of work on Mars already under his belt, Lowell traveled to the Arizona territory and chose a mountaintop just outside of Flagstaff to build his observatory. With a family fortune in textiles at his disposal, he built one of the finer

American observatories of its time atop the perch he renamed Mars Hill. It should be noted that, within academia and "professional" circles, Lowell (regardless of that Ivy League degree in mathematics), was considered by many an amateur astronomer, despite rigorous self-training and his fine observatory. Nonetheless, for the next fifteen years he made careful and detailed studies of Mars through his large telescope, resulting in thousands of drawings and three books: *Mars* (1895), *Mars and Its Canals* (1906), and *Mars as the Abode of Life* (1908).

It is worth mentioning that Lowell was one of the few "intelligent Martian" promoters who backed up his assertions with some generally solid scientific reasoning. He wrote convincingly of the relationship between Mars's age, its distance from the sun, and early planetary formation to support his ideas. Reading his books today, and understanding the sketchy evidence available to him and others of the time, one can enjoy the road map of his logic and feel the genuine passion of his notions. He was far from a tabloid journalist, despite the fantastic thoughts he put forth. For example, examine (and endure) the wordy excerpt that follows:

> . . . [T]he aspect of the lines is enough to put to rest all the theories of purely natural causation that have so far been advanced to account for them. This negation is to be found in the supernaturally regular appearance of the system, upon three distinct counts: first, the straightness of the lines; second, their individually uniform width; and, third, their systematic radiation from special points. . . . Physical processes never, so far as we know, end in producing perfectly regular results, that is, results in which irregularity is not also discernible. Disagreement amid conformity is the inevitable outcome of the many factors simultaneously at work. . . . That the lines form a system; that, instead of running anywhither, they join certain points to certain others, making thus, not a simple net-

> work, but one whose meshes connect centres directly with one another,—is striking at first sight, and loses none of its peculiarity on second thought. For the intrinsic improbability of such a state of things arising from purely natural causes becomes evident on a moment's consideration. . . . Their very aspect is such as to defy natural explanation, and to hint that in them we are regarding something other than the outcome of purely natural causes.[8]

In other words, these lines must be artificial canals because they are straight, uniformly wide, and go from one point to another, and this is unlikely to happen by natural accident.

Lowell went further, though, which may have done more to undermine his credibility than the faux canals. Another excerpt from the same book states: "Martian folk are possessed of inventions of which we have not dreamed, and with them electrophones and kinetoscopes are things of a bygone past, preserved

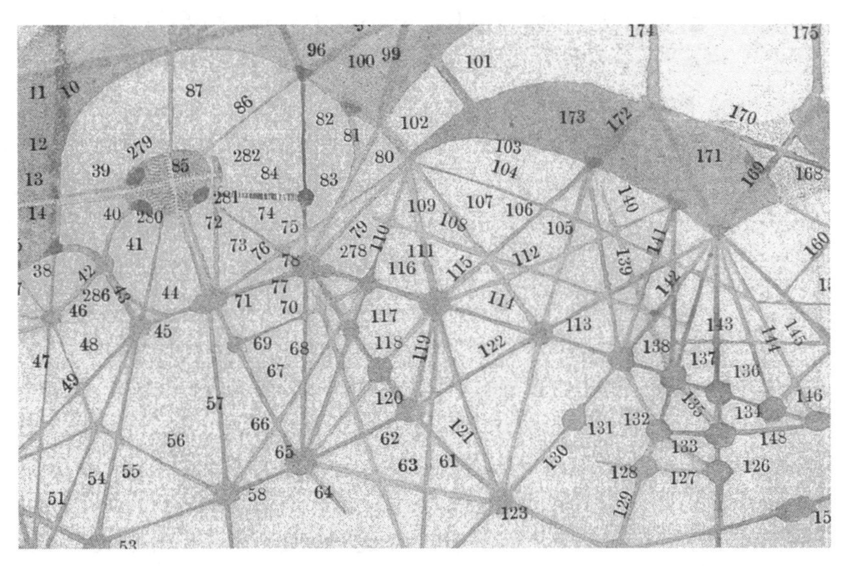

Figure 3.4. CANALS: Percival Lowell created extensive maps of Mars. Sitting night after night at his telescope, he would sketch the barest outline, then fill in details by memory later. His drawings were detailed and, some would say, imaginative. *Courtesy of NASA.*

with veneration in museums as relics of the clumsy contrivances of the simple childhood of their kind."[9]

Lowell, somewhat characteristically, took things a bit too far in his flawed but admirable enthusiasm. However, his copious writings, dated though they are, make for entertaining reading even today.

It has since been argued that Schiaparelli, Lowell, and the others who so patiently charted the lines across the Martian surface may have done little more than traced the capillaries in their own retinas, either as shadows cast through the structure of the eye or as reflected in the eyepieces of their telescopes. Nobody can be sure. What can be said with some authority is that no two observers saw quite the same patterns, and few users of modern telescopes have felt compelled to make note of such patterns in the last one hundred years.

Nevertheless, the stage had been set for a Mars peopled in some way, by some *thing*. Fiction responded to this fertile landscape via men like H. G. Wells, whose *The War of the Worlds*, first published as a serial in Britain's *Pearson's Magazine* in 1897 and later as a novel in 1898, wrote the first truly compelling tale of interplanetary invasion. His account, well-spiced by a background in journalism (and written in this fashion), was perhaps the first truly horrifying work of science fiction, in many ways (along with writers such as Jules Verne) establishing that genre. It was a newslike accounting of the invasion of Earth by Martians, and it remains as chilling today as the year it was first published. Navigating the void between worlds in meteorlike spacecraft, the aliens fell to Earth and dispatched three-legged machines of war as invincible as they were fantastic. And to make matters worse, they did not only lust for our planet, but further insulted humanity by feeding on our blood through small tubes inserted directly into their stomachs on one end, and into human bodies on the other. Wells was a known critic of the British penchant for colonial adventures and empire, and was, as later charted by academia, quite consciously commenting upon the decimation of less advanced societies by warlike technological powers such as Britain. He could not have done so in a more colorful fashion. The book remains in print today, over a century later, in dozens of languages.

Regardless of Wells's motives, *The War of the Worlds*, along with the ideas of Lowell, set the stage for popular thinking about Mars for almost seventy years. Adding to this was an offshoot of Wells's work, the 1938 radio dramatization of the book by Orson Welles (no relation) on his *Mercury Theater of the Air* radio program, which created a panic throughout the parts of the American East Coast within reach of the radio station.

No literary conversation about Mars would be complete without mention of Edgar Rice Burroughs. Prior to his most famous literary invention, Tarzan, Burroughs penned a deeply imaginative suite of tales about Mars. The planet was known as Barsoom to its inhabitants. And what inhabitants they were. Barsoom teemed with a multitude of beings (some red and some green), eight-legged fighting steeds, airships and castles, kings and queens, princes and princesses, and classic bad guys and good guys—all at war with one another. In the latter category fell John Carter, quite literally. The hero of the tales, Carter was a veteran of the US Civil War who was working in the US frontier in Indian territory. Losing his way, he found himself hunted by angry Apache warriors. Hiding in a sacred Apache-owned cave, he was mysteriously transported to Mars, where he was introduced to the fighting races of the planet and, more memorably, the barely clad Princess of Mars. He began a series of deeply involving adventures, eventually returning to Earth. If only NASA could find that cave in the Western states, its annual budget would go much, much further . . .

As late as the 1950s, an inhabited Mars still glowed brightly in the popular imagination due to a spate of motion pictures (notably, 1953's *The War of the Worlds* and *Angry Red Planet* in 1959) and fast-selling popular literature such as Ray Bradbury's *The Martian Chronicles* (1950). This last work was unusual in that it was a poetic and sympathetic look at the human colonization of Mars as a metaphor for the violent annexation of the Western frontier of the United States, complete with spiritually enlightened indigenous natives who are eventually wiped out by the bacterial plagues that hitchhike to Mars within the bodies of white men. Bradbury's take on Mars was not wholly dissimilar to Wells's in his observations of human malignancy.

Thus, the possibility of Mars as a place occupied by intelligent and possibly warlike beings remained deeply ingrained in the popular consciousness. And despite the opposition by many scientists of the day who declaimed these Martian fantasies due to the extreme temperatures and lack of measurable water or a breathable atmosphere there, this popular mindset lived on.

Then came 1965, and Mariner 4.

CHAPTER 4

THE END OF AN EMPIRE

MARINER 4

Mariner 4 represented NASA's first journey to Mars, the second of two spacecraft to attempt the trip. This was a time when the space agency wisely launched its unmanned probes in pairs, which often saved the mission. Mariner 3 failed, but Mariner 4 sailed past Mars, returning twenty-one spectacular, if ghostly, images. In the process, Mariner smashed the Martian empire of previous generations to, quite literally, dust. The images returned by that robotic craft were grainy and indistinct, and involved some sophisticated guesswork to interpret, but they showed a Mars that astronomers had scarcely imagined for all their years at the eyepieces of their mighty telescopes. For what those marvelous instruments could not reveal was the truth that emerged from the Mariner images: Mars was a desert, a place of craters and windswept stone, and vast fields of oxidized sand. It was, in short, an apparently dead world.

The mission was not an easy one. Launched with its twin, Mariner 3, only Mariner 4 made the full journey. Both craft departed Earth in November 1964, with Mariner 3 failing to fully jettison its protective shroud in Earth orbit. Unable to extend its solar-power panels, it soon died of battery starvation and now resides, mummified, in a lazy orbit around the sun.

Mariner 4 fared better, following the planned trajectory to Mars. It is worth relating that after observing the failure of the fiberglass shroud to release Mariner 3, technicians descended to the cape and, within a matter of a few weeks, designed a new nose cone from metal with an improved release mechanism. This quick fix was indicative of those rough-and-ready early days of space exploration, and the new design worked perfectly. Mariner 4 left Earth orbit without incident in December and headed off for the long voyage to Mars, fifty-four million miles distant—if one could fly in a straight line. But one could not, and the distance traveled would be far longer.

This was NASA's first success with Mars and only the second time a still-operating robotic probe flew by another planet (the Soviet Union had attempted as much, but all its machines failed before executing their primary missions). Mariner 2, launched in August 1962, had performed a successful flyby of Venus later that year (Mariner 1, true to form, was destroyed shortly after launch).[1]

Mariner 4 arrived at Mars in mid-July 1965, returning twenty-one good images and part of a twenty-second as it sped past the planet just over five thousand miles distant. The craft was crude by today's standards, but in 1964 terms it was a miracle of engineering for the unknown. In total, the basic instrumentation on the ship would send home twenty-three million scientific measurements. Not all of these were specific to Mars; many were measures of dust in space en route, attempts to measure both Earth's and Mars's magnetic fields, and many others.

Imaging was a paramount goal. The twelve-pound TV camera's images were stored to an onboard tape recorder and relayed back to Earth—twice to reduce the likelihood of errors—beginning a few hours after it left the Martian system. A primitive onboard computer converted the images to radio code to be relayed back to JPL via a global network of huge tracking dishes. Each of the twenty-one images, which were only 200 lines in resolution (the high-definition TV of today is 1,080 lines) took almost *nine hours* to download.

The operation of this camera was intended to be completely automated. There was a long delay between ground-sent commands, reception by Mariner, and a return confirmation, due to the extreme distance from Earth, so various sensors were installed to enable the system to operate autonomously. These were basic light-measuring systems that would sense when Mars was close enough to illuminate a photosensitive element to a certain level, then trigger the various parts of the imaging system as the craft swung by Mars. It was a one-shot deal, and any error could negate the entire trip. The cameras had to eject a lens-cap (controllers, concerned about a failure at this step, had accomplished this through a "cheat" in operations months prior). They then had to warm up and begin the picture-taking sequence right on time. Then, as Mariner 4 disappeared behind Mars after the flyby, the automated tape recorder (with three hundred feet of tape) would turn on and off in bursts to record the images for playback later, as the Earth was, for now, out of sight from Mariner's perspective. And this was just the imaging system—there were eight other experiments on board, most of which needed to perform in a timely fashion near Mars (one, an ionization chamber experiment to measure charged particles, had malfunctioned on the way from Earth).

Worthy of mention is the occultation experiment, simple in design but very effective. At the moment that Mariner 4 began to go "behind" Mars from Earth's perspective, a radio transmission was scheduled to allow JPL to measure the effect of the Martian atmosphere on the radio waves. This allowed for a basic measurement of atmospheric density, the first proper measure of this property. It was far, far lower than expected—about 4 millibars. Prior to this it was thought to be perhaps 30 millibars. This was just one more nail in the coffin of the old Mars of the popular imagination (Earth is about 1,000 millibars).

Many hours after the flyby, when the imaging playback had been collected at JPL, the data was fed through huge, state-of-the-art computers that were so slow that the imaging team took to posting the numerical printouts on the walls and using felt-tip markers or pencils to approximate the black-and-white value of the digits; they literally painted pixels. This gave them roughly shaded approximations of the imagery to enjoy until the computers had crunched through their tasks.

Between these and the later computer-built images, Mars was finally visible in all his martial glory. After centuries of blurry images through telescopes, always wavering and swimming in-and-out of focus at the whimsy of the Earth's turbulent atmosphere, the Mariner science team could finally see the surface of the Red Planet as it truly existed. It was a thrilling, game-changing moment.

The many craters visible were something of an epiphany to many Earth-bound observers. While it was suspected that the surface would be older than our planet's, the evidence of myriad craters, some huge, caused scientists to speculate that the planet's surface was not only very old but also relatively unaffected by erosion from weather or significant volcanic activity. Mars was declared, in general terms, a dead world. Later investigations would recant this to some degree; it is far from dead, though not lush by any stretch of the imagination. It has simply been far more active in weather and water terms than was thought at the time.

While it was stunning to see Mars looking much like Earth's own moon, not everybody was gratified. There were still many "Lowellians" out there insisting that the Mars they had held so dear in their mind's eye might still live. In short, they were not ready to let go. Ideas were hatched about underground civilizations, hidden structures, and subresolution indicators of advanced life (the Mariner 4 cameras were very low-res by modern standards). Only the later Mariner 9 and, finally, Viking probes would put these notions to rest for all but the most fanatical or asylum-bound.

But in an ironic twist, these folk were not all wrong. The low-resolution Mariner images showed a macroview of the Martian environment but were unable to resolve the more delicate and subtle features of this very complex planet. While large craters were visible, the gentle erosion of their rims by wind and sand was not evident. While broad plains showed up, the intense water-sculpted features were invisible. And so, while the canals were most certainly obliterated and the dying cities vanished without a trace, the chance for a once-warmer, wetter Mars, perhaps even hosting microbial life-forms, existed within the shadows of those twenty-one pictures.

One interesting characteristic of this flight was the attempted use of solar-wind steering augmentation. Each of the four solar panels, spread like an iron cross out to each side, had an additional "petal" extending from its tip, consisting of a skeletal frame covered with Mylar® foil. These outboard vanes were steerable and could be angled relative to the sun and the solar wind emitting from it. The idea was for these vanes to allow controllers to even out the pressure from the solar wind that might push the craft off course. While slightly effective, such steering-vanes were not used again on a Mariner flight.

One surprising discovery was from Mariner's attempt to measure Mars's magnetic field, which ended with mixed results. The primary finding was that it was unexpectedly weak. Only later would the true—and surprising—nature of this be understood.

Recall that this was the United States' first foray to the outer solar system, and much more was unknown than known. Cosmic rays were measured, and another instrument attempted to track the number of collisions with material in space—hopefully minute in size—and did in fact record a few hundred hits, all with tiny particles of dust. Space, at least within the inner solar system, was far from "empty."

Mariner 4 was a success by any standard. In fact, it operated for many months after its encounter with Mars, and could have sent back data far longer. But in late 1967, Mariner encountered an area of space dense with dust and grains from an extinct comet, and suffered dozens of violent impacts of varying sizes and intensities. Not only did these impacts appear to have damaged the delicate craft, but they also seemed to push it off course. Soon thereafter it ran out of maneuvering fuel, and JPL switched off the machine. It eventually joined its sibling, Mariner 3, dead and still ensnared in its faulty launch shroud, in a large orbit around the sun.

The few images Mariner 4 returned showed Mars to be a dry, barren place, festooned with craters. The mission was a changer of worlds, and the Mars of Percival Lowell and H. G. Wells was forever banished. The canals and fighting machines that had dominated the first half of the twentieth century were relegated to their proper place in the imagination, left to flower for a new generation of science fiction readers, who had to balance their passions with the stark realization that a Mars peopled with intelligent life, whether friend or foe, was no more.

CHAPTER 5

DR. ROBERT LEIGHTON

THE EYES OF MARINER 4

In the world of 1964, there were few go-to people when it came to exploring the solar system with robots, especially when it came to actually *seeing* the planet below. At this time, Russia and America had launched numerous satellites into Earth orbit and flown probes to the moon. But to other planets? The Soviets had attempted Mars six times, but all these missions failed, as did their three attempts to reconnoiter Venus. The United States had succeeded with a mission to Venus in 1962 with Mariner 2. But we must remember that the first Earth-orbiting satellite had been launched only in 1957, so the space age was young.

It was therefore a daring feat when NASA detailed JPL to send twin probes to Mars, Mariners 3 and 4, in 1964. One goal was to transmit TV images back to Earth, both for scientific and political reasons. The political motivations are obvious: we were in a space race with the Soviet Union, a battle of political systems and technological might. While this was paramount with the manned space programs, it had begun with Sputnik and Explorer (the first Soviet and American satellites, respectively), so unmanned exploration was also an important measure of success as well. The scientific ones were also relevant: prior to the success of Mariner 4, Mars had been seen through Earth telescopes in some detail since the mid-1800s, but viewing through the syrupy atmosphere of Earth provided little in optical detail. Mariner 4 had the potential to convert the hazy, fuzzy telescopic images of Mars into relatively crisp, clear images of the planet most like Earth. It was time for the next step.

When JPL set out to build a team to design an onboard camera, one of the heavy hitters it sought out was a Caltech professor named Robert Leighton. Born in 1917, Leighton had attained a bachelor's degree, a master's degree, and a doctorate from Caltech and stayed on as a professor. Leighton was a garrulous fellow, plain-spoken and capable, and never hesitant to speak his mind in easily understood terms. Trained as a particle physicist under the likes of Richard Feynman, his interests ranged wide and far, and he was fascinated by this new and exciting venture. The passion he brought to the program was infectious and helped to make Mariner 4 the great success it was. He was named the principal investigator for the TV-imaging experiment.

Dr. Leighton, who died in 1997, was interviewed numerous times about his ventures into space.[1]

"In '61 or '62, I was dragooned by Bruce Murray and Gerry Neugebauer into participating in the Mariner 4 photographic experiment, [the] television experiment, and the reason for that was that there had been no reasonable proposals for a photographic component of the mission on television, for pictorial work. I think people had made studies for NASA of camera systems and stuff like that, some written stuff was available. But nobody had proposed, for that mission, putting together a particular kind of a TV camera."

Leighton and his team would soon change this. Integral to this project was returning images of Mars; at this point, the romanticized world of Percival Lowell, while extremely unlikely, still could not be entirely ruled out, at least not in laymen's minds. But when the haunting black-and-white images from Mariner 4 did come through, everything changed.

"The reality of it just suddenly waked everybody up. Percival Lowell left us with all that business, 'waves of darkening' and things. And unfortunately . . . some of the members of the scientific team are basically romanticists. You know, they are just unbridled romantics . . . 'if you can't prove that there isn't life on Mars—well, then there must be life on Mars, and let's go find it.' Well, [Mars changed]. It's got dust storms and polar caps and things.

"It's obvious . . . there'd be craters and everything. And yet, the fact that craters were there, and were a predominant land form, was somehow surprising. And the [imaging] limitations were so severe . . . that we waited a week or more, after we knew there were craters, before any kind of official announcement was made. At JPL things were protected very much, because it's one of those things where, if somebody had leaked [the term] "craters" . . . I think what we were trying to avoid was being drawn into a detailed discussion of things before we had had a chance to make any kind of measurements . . . so we took a week or two, and made measurements, and then had a press conference."

Though a physicist by training, Dr. Leighton had become a respected astronomer over the years, specializing in planetary photography. His approach to the needs of the Mariner flights was interesting in this regard.

"Say you wanted to devote a certain amount of money, over a period of time, including mainly space experiments . . . as a scientific component of the space program, you want to bring back the most science within the area of coverage that you can for the amount of money [you have].

"I think time and again, the atmospheric pressure on Mars, the water vapor on Mars, the temperature of Venus—and there's other things—the rings of Uranus and so on, have come first, or at least equally from ground-based work. And [while] the space work [was] a unique contribution . . . the prior knowledge of those properties, if we'd known them a few years earlier, could have greatly enhanced the scientific return from the missions that were flown. And so I've always felt that one of the first things NASA should have done was to build four more 200-inch telescopes . . . as it was, one mission—one damned spacecraft mission would cost as much as five 200-inch telescopes, plus the mountains to go with them."

He felt that more could be done, once in space, by furthering the knowledge that could be gained from Earth via telescopic observation. However, at this point, imaging from Mars was becoming critical to Mariner 4. But to date, the approach seemed to him to have been somewhat ad hoc:

"We were sort of approached by JPL. The people there had done a lot of thinking about [imaging], but they didn't have any scientists. They had the technical know-how, and had the tubes and everything else, and they'd even sent the Rangers [to the moon] . . . they had good cameras. And [JPL] had a lot of experience with television cameras and so forth, and to the extent that they thought they were just going to the moon again, they were well up the curve. And so by the time they latched onto a few of the scientists, and we got together and made a group that would do it, it just was a leaderless, headless thing, where there was knowledge . . . but it was not in places where the people . . . could propose as a team. So it was a sort of a fluke, in a way."

So, in the end, was it all worth it? On this point, he seemed assured.

"Oh yes, absolutely. And that's one of the best parts of it all. Some of the letters that came in, from the milkmen, the dairy farmers in Oregon—they'd been watching TV at, I don't know, 5 a.m. or whenever the thing went over [the airwaves] . . . they said, 'I'm not very close to your world, but I really appreciate it, keep it going.' I thought that was kind of nice."

It was indeed. And it was the beginning of a whole new understanding of Mars.

CHAPTER 6

CONTINUING TRAVELS TO DARK AND SCARY PLACES

MARINERS 6 AND 7

In 1969, Mariners 6 and 7 continued the quest by performing two more flybys of Mars, this time at a range of only two thousand miles from the planet instead of five thousand as with Mariner 4. And both spacecraft made the trip successfully.

While these probes were bigger and heavier than Mariner 4, they were assigned a similar mission: fly past the planet and perform as much science and imaging as possible. By following divergent paths past the planet, the twins would cover as much of Mars as possible while whizzing past. It was, like Mariner 4, a bit like firing the spacecraft out of a cannon, at a point where Mars would be when they reached its orbit after a 200-million-mile voyage, and snapping photos as they passed it. It was a carefully controlled train wreck, with one train snapping pictures of the other as they crossed paths.

Complex as it was, there was more drama behind the scenes. While this Mariner mission was one of the few in this era in which both twins would complete the voyage together, Mariner 7 nearly didn't make it. A week prior to encountering Mars, the craft dropped out of Earth contact. JPL mobilized its Deep Space Network and began an intensive effort to make contact with the missing probe. After numerous efforts, contact was regained via the low-gain antenna, and the craft was able to reorient itself and resume the mission with only minor delays. While at the time it was suspected that Mariner might have been impacted by a meteorite, it was later decided, based on the best sleuthing JPL could manage, that an onboard battery had exploded, temporarily disorienting the craft. In the final analysis, Mariner 7 actually outperformed its twin during the high-speed flyby.

The mission of each craft, while similar, had slightly differing objectives. Since Mariner 6 launched before Mariner 7, and the Mars encounters were separated by almost a week, the data gleaned from 6 allowed mission planners to implement last-minute changes to the data-gathering efforts of 7. Each had new and improved camera systems, one for wide-angle imaging and one for telescopic imaging. This was a vast improvement over Mariner 4, with its single low-res camera.

In addition, each craft sported an infrared spectroscope, an infrared radiometer, and an ultraviolet spectroscope. These instruments were improvements over the Mariner 4 package, and a reaction to the discoveries made by that pathfinding mission. By today's standards, this was a very basic package, but for the time it was quite sophisticated and incorporated a lot of lessons from Mariner 4.

Unlike that first flight, these two Mariners were not tasked with performing scientific investigations during their cruise to Mars. Their specific goal was to begin—and end—their primary operations at Mars encounter. The space age was maturing. They were to fly over the equator and southern hemisphere of Mars, respectively.

Each craft was just under one thousand pounds in mass (or about double the mass of Mariner 4), and measured ten feet tall and about nineteen feet wide when in cruise mode. As before, there was an analog tape recorder for image storage and retransmission. There was also a digital recorder for science data storage. And, like Mariner 4, internal temperature control of the spacecraft, critical to successful operation, was governed by an ingenious system of slatted louvers, like Venetian blinds, on the sides of the boxy main body.

One stunning difference was the amount of data that could be sent back to Earth. The designers of Mariner 4 had been very conservative, so concerned were they that things be kept simple to ensure mission success. With these new spacecraft, there was an intense effort to maximize the data return and speed with new techniques; but this was not implemented without conflict. Conservative engineers sparred with the more aggressive science team in a running battle over data rates. Ultimately, the newer system was couched as an "engineering experiment," which would have provided political cover in case of a failure. But the net result was that its data transmission was about two thousand times that of its predecessor. While a stunning success, this would not be the last time unmanned space exploration would suffer internal squabbles over political concerns.

Launched a month apart, the twin spacecraft arrived at Mars separated by only five days. Mariner 6 entered the vicinity of Mars on July 29, 1969. The probe was flying via inertial guidance as opposed to relying on the problematic Canopus sensor. This had caused much drama on Mariner 4. And on this flight, when the explosive retainers of Mariner 6's scan platform were ignited to release the cameras and instruments into their postlaunch science-gathering configuration, they blew a small cloud of detritus into the craft's immediate vicinity. Of course, being in midflight, the particles kept right on traveling with the spacecraft, and the Canopus star sensor began trying to lock on these small, brightly lit particles instead of the proper guide star. So rather than track bits of paint and ash as they flew past Mars (and possibly end up imaging the black of space instead of the planet), they relied on the internal guidance system to orient the craft.

Fifty hours prior to flyby, the instruments were activated. Two hours later, the "bomb run" began, and for the next forty-one hours, forty-nine images were snapped of the approach to Mars via the narrow-angle (or close-up) camera. On July 31, the close encounter began, and another twenty-six close-ups were captured.

The craft sped past Mars at a distance of about two thousand miles, or less than half that of Mariner 4. As it left Martian space, it sent home the data and images it had acquired, then began a limited set of observations of the outer solar system and the Milky Way's galactic edge.

The only equipment failure on Mariner 6 was insufficient cooling of the infrared spectrometer at Mars encounter, limiting its usefulness. Overall, though, the mission was a crowning success. Shortly it would be Mariner 7's turn to barrel its way past Mars.

Despite the early navigation and communication problems, Mariner 7 behaved very politely as it neared its target. Reaching Martian space on August 2, 1969, the craft captured ninety-three images of Mars as it closed on the planet. After fifty-seven hours of approach, the close flyby began. The craft was taking a more southerly trajectory than originally planned, due to early analysis of Mariner 6 images. There was some awfully interesting scenery showing up in the southern hemisphere, and this was a chance to see it close-up. By August 5, the craft had snapped an additional thirty-three close-encounter images and left Mars. As it sped away, it followed a similar program to Mariner 6's, adding observations of a small comet to its dance card.

After the first tantalizing twenty-plus low-res images from Mariner 4, Mariners 6 and 7 were a smash. A total of 201 images were sent home, with fifty-nine of these being close-encounter images. These images covered about 20 percent of the Martian surface. Ultraviolet and infrared data were acquired from the planet, and the atmospheric pressure was further refined to about 7 millibars. The south polar cap revealed itself to be composed primarily of carbon dioxide after all (a thick layer of water ice would be identified below this over thirty years later).

The missions sent back enough information to allow for future planning of missions like Mariners 8 and 9, as well as early thinking about the Viking mission. Surface composition, atmospheric density, and ambient temperatures were tracked and

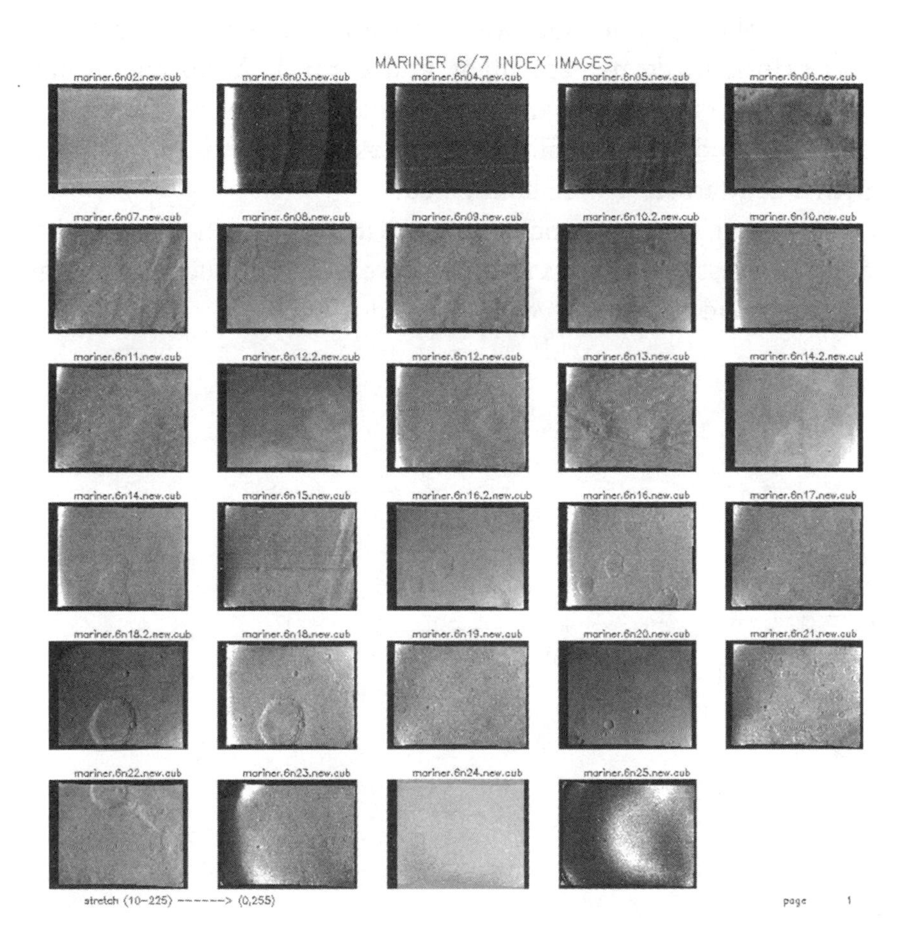

Figure 6.1. MARINER REDUX: The missions of Mariner 6 and 7 were a replay of Mariner 4, but with everything done bigger and better. The craft were larger and more capable, they flew half again as close past the planet, and took many more images. This index of near-encounter images shows a Mars that is still arid in appearance, but some evidence for weathering is starting to appear. *Courtesy of NASA/JPL.*

studied. The mass of Mars was refined and lots more experience was gained in deep-space flight and control. Atmospheric data showed large amounts of dust present, as well as water and carbon dioxide clouds, and finally, the atmospheric pressure measurements of Mariner 4 were confirmed.

All in all, these Mariners accomplished about all that could be done in a fast flyby. It was time to try something much more challenging: send a craft into orbit around Mars.

CHAPTER 7

DR. BRUCE MURRAY

IT'S ALL ABOUT THE IMAGE

In his own words, Bruce Murray was "drafted . . . reluctantly!" to work on Mariner 4 fresh out of grad school, but in the end, his collaboration with people like Robert Leighton resulted in not only a stellar career but also a fascinating path through deep-space exploration.[1] Arriving at Caltech as a young man, he blossomed within the Mariner program and eventually headed off to spearhead Mariner 10 to Mercury—a single-spacecraft, low-cost, and high-risk mission that would be a precursor of "faster, better, cheaper" at JPL. In time he became the director of JPL, then retired to continue evaluating the massive data dumps from the Viking missions that had been stored and largely uninvestigated. Never a shy man, Murray is not one to mince words, and when he takes on a fight, his opponents had best be prepared to defend themselves . . . even if that opposition is NASA. Such an event occurred early in his career, as Mariners 6 and 7 were being designed.

"Mariners 6 and 7 were to be carbon copies of Mariner 4, and we had a long struggle with JPL and NASA to upgrade them with greater information capacity. This was the principal theme I played: to increase the amount of data to be returned, which was easy to do. For example, on Mariner 4 . . . the telecommunications channel to return the data . . . only returned eight bits per second. That's like a teletype . . . It could have returned much more, and Mariners 6 and 7 would have been a similarly conservative design. But I, in particular, had a real battle with them. I had to learn about the electrical engineering, I had to learn about the communications, I had to figure out all this stuff. I realized what was happening was that the engineers were protecting themselves. They put tremendous safety margins in so that if anything went wrong they would still be able to get the signals back. But they paid for that with poorer signals. By choosing to get one-hundredth of what they *could* get back, they were much more certain of getting it done. It was really a sociological and psychological problem once you understood the technical principles. And that led to a long battle. Mariners 6 and 7, in fact, had sixteen thousand bits per second instead of eight bits per second. It was a factor of increase of two thousand, as a so-called engineering experiment."

The term *engineering experiment* was accurate to a point but obscured a larger truth in Murray's mind: it would cover the engineer's back ends if something went wrong. He was not the type of person to tolerate that kind of game. He preferred to aim for the highest performance possible and to take your lumps if it didn't work.

"In all this, I taught myself other things. I wrote some papers . . . about communications to establish that I knew the subject better than they did. One of the lessons I learned from this was that in order to be successful, the scientists had to learn the engineering at least as well as the engineers; I had to learn about spacecraft stabilization, had to learn about power, had to learn about active control, had to learn about scan-platform motions. Television was the big drive, all the way through. One of the reasons I was able to become the director of JPL was that in order to do that job I had to understand the whole spacecraft."

But it all began with Mariner 4, and the follow-on missions of 6 and 7 had to be different in some meaningful way.

"So the Mariners 6 and 7 mission: one was targeted in equatorial areas and one was to go at higher latitude over the polar ones. Mariner 7, which had a lot of technical difficulty, such as a battery explosion a week before encounter, was in fact able to confirm that indeed the seasonal caps are CO_2 . . . that was a major discovery. [The Mariners] discovered some collapsed terrain. [They] discovered some other physiographic features, but [they] didn't discover either channels or volcanism, which is, again, how Murphy's law operates in science. We have to look at the wrong places [first]. So it was not until Mariner 9 that the most significant surface features—these huge volcanic structures, and the huge water-carved channels—were discovered."

But no probe can look *anywhere* if it can't talk to mission control. During the early space age, tracking of spacecraft was a rather ad hoc affair. One of JPL's crowning achievements in planetary exploration was the Deep Space Network, or DSN.

"The Deep Space Network is one of the most marvelous products of American science and technology in the world. It is also a marvelous attribute of JPL, and to a lesser extent of Caltech. . . . [It] consists of three principal stations for commanding—that is, communicating to and listening from spacecraft that are at some great distance from Earth. They are located at approximately 120-degree longitude differences, so as the Earth turns, the spacecraft is always in view of one of the tracking sites, which can spot occasional problems. There used to be one in South Africa, which was good, because it gave southern-hemisphere coverage and also gave additional longitude, which was about the same as Madrid. That was closed because of political pressure on US–South African ties. Among the many minor histories of JPL, there was a short period there where the official position was that we were out of there, but in fact we were still in collaboration. One can argue that sort of thing both ways. We also had to deal with Franco of Spain. . . . Institutionally, this has been operated, developed by JPL exclusively with relatively great independence from NASA. . . .

"Technically the JPL group has been so superior and so outstanding, superior to anybody else in the world, that they really have been much more in charge of their own destiny than others. . . . The Deep Space net has gotten very, very good at introducing new technology into an ongoing, highly reliable system. It's a very, very impressive situation. They've also been able to structure the administrative and political arrangements so that they can plan ahead. The plans get changed, but they always have a long-range plan. They're always developing new technology. . . . It's an integrated system, sort of womb to tomb. . . . The result of this is they've been able to use state-of-the-art technology consistently and bring it in, and that net has been superior to anything else in the world, almost from the time it started, and it just gets better and better and better. It's absolutely incredible. It's a magnificent thing.

"The fact of the matter is they have been absolutely outstanding, and this, coupled with a group not part of them but part of the regular JPL, which does the navigation—takes this information and converts it into location and trajectories and these sorts of things—have been truly outstanding. There's just nothing anywhere close in the world."

Planetary exploration was to suffer cuts and inconsistent funding over the next forty years, but one area would always get the money needed to continue operations.

"[The DSN] has shown remarkable resilience. Even when there are no missions, NASA keeps letting more money be put in, because NASA can never say there will never be any new missions. They can always say, 'We won't make a new start this year.' But for the tracking people it goes on. So that's moved marvelously well."

Returning to the missions themselves, Murray recalls what he felt was probably the most important part of the flights past Mars, the most vital component.

"The imaging, of course, was of greatest public value. The reason is an important one: namely, that you don't need an interpreter for pictures. [For] everything else, you have to have the scientific priest interpreting the Bible for you. With pictures, however, the media are set up to transmit them—newspapers, magazines, or television—all over the world. We've developed it to a high art. . . . Bang! It just goes out! So I would argue that the positive aspect of the pictures was that it let people participate in the exploration, and that's very good. But the result is that the NASA administrators—in general, not always, but certainly the JPL project people—were very pro-imaging, because they saw recognition coming from this. There's a lot of positive feedback, and they felt it."

It's hard to imagine the exploration of Mars proceeding without all the wonderful images that have come back over the decades, but without people like Bruce Murray, that might have been the case. For this, and the proper utilization of the "scientific priests," we thank him.

CHAPTER 8

AEOLIAN ARMAGEDDON

MARINER 9

As the 1960s churned onward, the lion's share of NASA's budgets were devoured by the omnivorous Apollo program to land on the moon. Driven by politics as much as by science, Apollo was an insanely complex undertaking, and by the time Mariner 4 had departed Mars for the deeper solar system, Apollo had already reached the peak of its spending. Never again would NASA command so much of the national budget (at its highest, in 1965, about 5 percent). JPL and the unmanned exploration of the solar system had to make do with the leftovers.

But what a job the lab did with the money at hand. At the time of Mariner 4, JPL was also flying the Ranger series of missions to the moon. And the Rangers were traveling *to* the moon—in fact, they slammed smack into it, as planned. This was a simple, if inelegant, way to get ever closer images of the lunar surface. Orbital probes could do wonders from above, but even the most sophisticated of cameras had resolution limits. But by flying the probe right into the terrain, the final stream of images gave an ever-better view of the rocky, forbidding surface. It just had to be gathered quickly, *very* quickly.

After the Ranger program was completed, and after the flight of Mariner 4, the Surveyor program targeted the moon once again. These were landers, and infinitely more complex than the Ranger spacecraft. They had to navigate, land, and perform surface tests on the moon. And of course, with Apollo so dominating the national space agenda, they had priority over Mars missions, for the results of the Surveyor probes would help to design, and pick landing sites for, the Apollo lunar modules. Nonetheless, JPL had a knockout Mars mission planned for 1971: Mariner 9.

As a continuation of the Mariner series to Mars, it had a similar configuration as its forebears. But Mariners 8 and 9 each weighed as much as two of the previous ships, mainly due to the tremendously heavy extra fuel required to brake into Martian orbit. This fuel load made up almost half of the spacecraft's mass.

Of course, as with all the Mariners to date, a twin spacecraft had been part of the flight manifest. Mariner 8 launched in May but failed early in flight and ended its mission by splashing into the Atlantic Ocean. Some at NASA and JPL must have been wondering if they should simply skip the first launch of a Mariner mission by now.

But its twin, Mariner 9, launched successfully three weeks later. After an uneventful cruise to Mars, on November 14, 1971, Mariner 9 arrived at the Red Planet. It was the first spacecraft to enter orbit around another world. By this time, though, the mission goals had changed. With the loss of its sibling, Mariner 9 had to pick up some of the potential lost when Mariner 8 went diving into the Atlantic. Mariner 8 had been the designated mapmaker, while Mariner 9 was to study atmospheric changes over time. So, through a clever sharing of resources, Mariner 9 aimed to combine both mission profiles to the extent that it could. Some compromises were inevitable; the imaging of the polar areas, for example, was done from a high angle of inclination, and the shots were therefore somewhat degraded. Despite these compromises, the mission would be a stunner.

Mariner 9's engine fired at just the right moment to slow the craft sufficiently to allow Martian gravity to grab hold of the robot and pull it around the planet. It was now in an orbit that dipped below one thousand miles—half the previous flyby altitudes—with a vastly improved imaging capability of about 320 feet per pixel (the previous probes had managed about a half mile per pixel). This was a huge leap over previous efforts. The craft was in position to fulfill its primary mission: image and map the planet below. Its builders were elated . . . almost. For this time, Mars, not the spacecraft, had a problem.

You see, the probe had arrived at a costume party uninvited. Mars was wearing a planet-wide robe of dust, the largest storm ever observed. Virtually the entire planet was enveloped in a solid cloud of dust. This was a fact of life . . . but the question now was: how long would the storm last? Days? Weeks? In fact, it was mid-January 1972 before productive imaging of the surface could begin, two months after the robot's arrival.

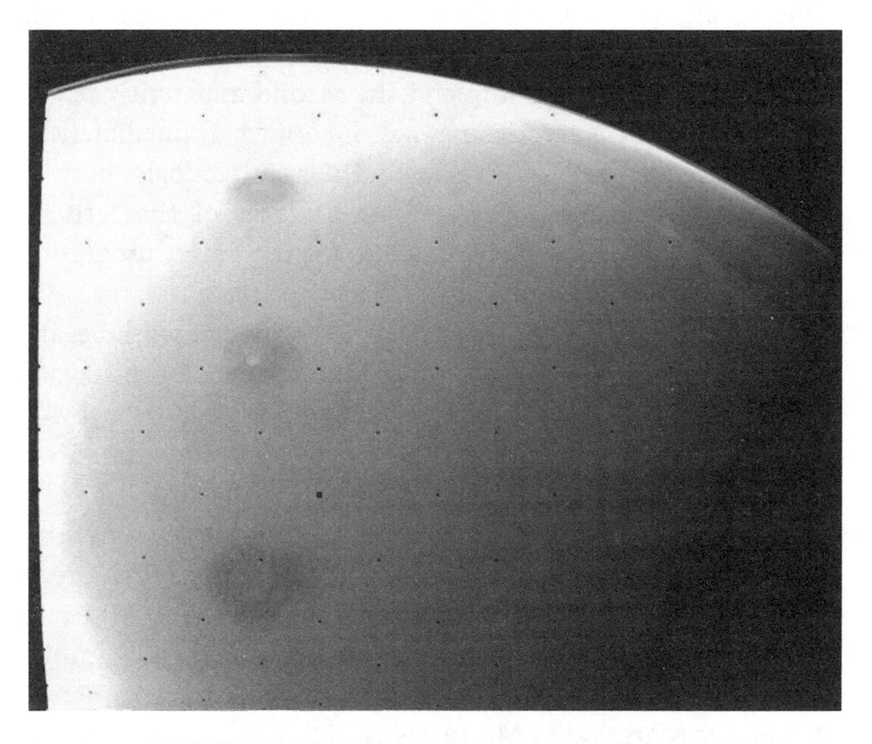

Figure 8.1. ATOP THE STORM: While the dust storm that enveloped Mars upon the arrival of Mariner 9 slowly waned, a few features became visible poking through the clouds. It was soon evident: these were the tops of three truly enormous volcanoes. *Courtesy of NASA/JPL.*

But the waiting period was not wasted. After a time, a trio of specks were spotted in the clouds below, poking up from the northern hemisphere. It was soon realized that these were the tops of mountains, and the mountains were volcanoes. They were identified as the three major features of the area known as the Tharsis Bulge. A fourth peak was also spotted, that being the massive top of Olympus Mons.

It is worth mentioning that since this was the height of the space race, the Soviet Union had dispatched its own set of twin spacecraft to Mars, dubbed Mars 2 and Mars 3. Both reached Mars shortly after Mariner 9. Unfortunately, these Soviet ships were not reprogrammable, as was the case with Mariner 9, and rather than wait out the dust storm, they proceeded to follow their programming right on schedule. Landers were dispatched from each, the first crashing and the second apparently reaching the surface intact but losing radio contact immediately. The orbiters fared little better; following their simple logic, both used up their available resources snapping images of the featureless dust clouds below. Another Mars loss for the Soviet program and another blow to Soviet prestige in the space race.

Once the dust storm abated, Mariner 9's real work began. In just under a year of operation, Mariner 9 mapped the surface of the planet in 7,329 dazzling images.[1] Previous Mars flybys had returned less than one thousand images, so the increase in coverage was nothing short of astounding.

But the real gems lay in the unexpected bonanza of surface features below, specifically the weathered terrain. The previous Mariners had sent back data from what appeared to be a mostly dead planet. Not so for Mariner 9. Gullies, streambeds, canyons, even clouds and fogbanks were seen. And of course, the incredible gash across Mars, Valles Marineris (which was named in honor of the mission), was discovered. The complex terrain patterns that crisscrossed the planet were charted. The massive volcanoes, seen in tantalizing glimpses through the dust storms of just a few months before, were extensively photographed. The unusual gravitational field, lumpy and surprising, was first measured by Mariner 9. Even the tiny, elusive moons of Mars were imaged. It was a hands-down winner of a mission.

Of these finds, the most arresting was the profound evidence of weathering. At the imaging resolution available, it was difficult to define exactly what they were seeing, but it was clear that the surface below had been heavily influenced by wind, water, or both. The extensive aeolian (wind-sculpted) features made sense; the dust storm that greeted the probe indicated as much. But what about the more robust areas of surface change? How had the channels been dug? Why were there areas that looked like river deltas? It was hard to ignore the fact that some of the weathered regions looked just like water-worn features on Mariner's home planet. But to jump to this conclusion meant that sometime, somewhere, Mars had enjoyed ample amounts of water. When? Where was it now? The general understanding of Mars, just a few years old at this point, stated that liquid water could not exist on the surface. So these enormous aqueous floods, which scoured areas the size of some US states, must have occurred well in the past. But what mechanisms were at work? As with every encounter with Mars, more questions were raised than answers supplied. But space scientists would not have it any other way. These were exciting times.

By October 1972, the attitude-control gas onboard had been depleted, and the mission was terminated with the spacecraft being shut down. Its mission was complete. Mariner 9 continues to orbit Mars to this day, sailing around the planet deaf and dumb in the cold darkness.

The legacy of Mariner 9 was to prepare the way for future missions, especially the Viking landers. The treasure trove of images was unparalleled for the time. The mission longevity was likewise impressive, and the flexibility of the flight, both in the recovering of the primary mission objectives once Mariner 8 was lost and the long stand-down upon the probe's arrival at a dust-enshrouded Mars, is legendary. It is a testament to the mission planners, those who executed it, and of course the spacecraft itself. This next-to-last Mariner (Mariner 10 flew past Mercury) was a demonstration of what a bargain these early missions were. For a total cost of $554 million, the inner solar system had been opened, and brilliantly.

Up next: Mars, prepare yourself to be invaded . . . by a Viking.

CHAPTER 9

DR. LAURENCE SODERBLOM

THE EYES OF MARINER 9

Laurence Soderblom got his first taste of planetary science working on the Mariners 6 and 7 mission. This was a time when *many* people were getting their first taste of planetary exploration . . . the field was only a few years past the age of the telescope. But Soderblom was bitten by the planetary bug early in life.

"As a kid I was interested in astronomy, and I actually built a spectrograph. I was in high school then, and my mother in particular was an avid rock collector and we used to go out to look for rocks and minerals a lot of the time. My dad's background was physical science, so when I went to college, I went into geology, and then into physics, and by the time I got done, I ended up with two complete bachelor's degrees in physics and geology. I went to Caltech, and planetary exploration seemed like a natural blend of physics, math, and geology."[1]

As a graduate student at Caltech in the early 1960s, Soderblom had the good fortune of landing Dr. Bruce Murray as his graduate advisor. While Dr. Murray may not have been the gentlest soul to ever grace that august campus, he was a forthright, hard-charging explorer of the cosmos. Soderblom felt fortunate to know him.

One thing led to another, and as the fraternity of planetary explorers is a close one, Soderblom's next stop was the United States Geological Survey. "I actually joined the USGS in 1970 when I finished my doctor's degree, and I went to work for Hal Mazursky, [who] was the lead of the television experiment on

Mariner 9, [which was] the first spacecraft to orbit another planet. We ended up mapping the entire surface."

That accomplishment was not an easy one even under ideal circumstances. When Mariner 8 was lost, it became much more difficult. "It was a scramble, because [Mariner 8] was [to create] a global map, to systemically map the planet. [Mariner 9] was an orbit that allowed repeated coverage of particular areas, to look for changes or activity, and the two orbits were quite different so they had to be blended together to satisfy both goals."

Of course, as Mariner 9 neared Mars, it was clear that the mission of the remaining spacecraft had just become still more challenging . . . the planet was enshrouded in a dust storm of global proportions.

"As a matter of fact that [dust storm] was kind of fun, because I noticed very subtle darkish patches in the cloud. I worked with the people at image processing and we invented a special filter, [because] we just couldn't see anything. So we ran [the filter] over the images and out popped four gigantic volcanoes! We knew [they] had to be really tall, to poke up through the dust cloud. The Martian scale height is about 30 kilometers [18 miles] and the dust, if it's mixed uniformly, which would be in the lower atmosphere, wouldn't drop to 30 percent of your level. So you get to thirty thousand feet, it had to be quite high. They were big, big volcanoes."

Eventually the storm abated, but it took time for the view to reach its full magnificence. "It clears gradually over many months; [the dust] takes a long time to clear the atmosphere. When it did, we started to see contrast in the polar regions, where there was ice on the surface that would stand off, and started to sublime and clean up. Of course, it never clears up completely, but it clears as much as it *ever* clears. There's always haze in Mars's atmosphere. But [the storm] cleared to the point that we could see as much detail as the camera would permit."

Mariner 9 replaced the few images sent back by the rapid flybys of the past with a year of orbital photography, and some surprises were inevitable. "We realized what Mariners 4 and 6 had suggested to us was that Mars was heavily cratered with an ancient surface. [When Mariner 9 arrived], and as the dust cleared, the channels, all of the fissures and streaks, the interpretation had swerved from being a very ancient dead Mars back to an exciting and complex, albeit desert, environment. That was the big realization. It was quickly apparent that there was likely flowing water on Mars, and that was in some of the Mariner 9 data.

"It was then that we understood: Mars was not just the moon painted red."

And perhaps that is the most elegant way of summarizing the profound nature of the Mariner 9 mission. It was a game changer. For while the previous Mariners had given us our first glimpses of this mysterious world, Mariner 9 gave us its true face . . . and what a face it was. The framework was now in place for a great leap forward—a robotic landing on Mars.

CHAPTER 10

VIKING'S SEARCH FOR LIFE

WHERE ARE THE MICROBES?

Viking had been planned from the start to search for extraterrestrial life—in this case, Martian microbial life. Toward this end, and after long and rigorous debate, a suite of investigative and highly portable experiments had been designed. Ingenious in their overall concept, simplicity, and execution, the suite was robust yet straightforward in its design.

The life-science experiment could be thought of as a man locked in a windowless room. We, as outside observers, have no idea if anyone is inside. This room may have no windows and no doors, but it does have a fancy mass spectrometer affixed to the only air exit from the room (bear with me here), and we are monitoring the mass spectrometer. The man eats some Italian food, heavy on the garlic. After he finishes the meal, give it a few hours . . . he might even pop a few Tums® . . . but eventually, he will probably emit a satisfied burp. Since the room is sealed, eventually the vapor from his emissions will make its way to the spectrometer. It will sense the compounds in his gaseous outpouring, and via this we will know that there is a man in the room, and have some idea of what he ate (and then we'd best let him out, as the air is getting mighty thin and smells like garlic!).

This, in simplistic terms, is how the Viking life-science experiment was intended to work. In reality it was much more complex. Now, imagine that a group of such rooms aboard the small spacecraft: the supply of Italian food and the instrumentation for four different experiments had to be reduced to a package far smaller than a cubic yard and weigh less than thirty-five pounds. That was the challenge facing the designers of Earth's first flying life-sciences lab.

But before they could utilize their amazing machine, they had to decide where to land. Given that their only resources to date were fuzzy telescopic images and data from Mariners 4, 6, 7, and 9, they did not have a lot to go on. So the drama of landing-site selection is a story unto itself, for nobody wanted to be the one person responsible for sending this billion-dollar mission to a lander-wrecking Martian rock quarry.

Mariner 4 had imaged scarcely 1 percent of the Martian surface . . . 6 and 7 had added to that, performing detailed mapping of about 20 percent of the surface. But there were still vast regions of terra incognita (or perhaps more properly, *Ares incognita*) unseen and unmapped. Then Mariner 9 covered most of the planet at some level of quality, but mission planners could be not be entirely sure just what they were seeing, as the resolution was not very high; anything smaller than half a mile in size was invisible. It would be easy to miss a threat the size of a small city block: a huge crater, an enormous outcrop, or a rock-strewn field. And

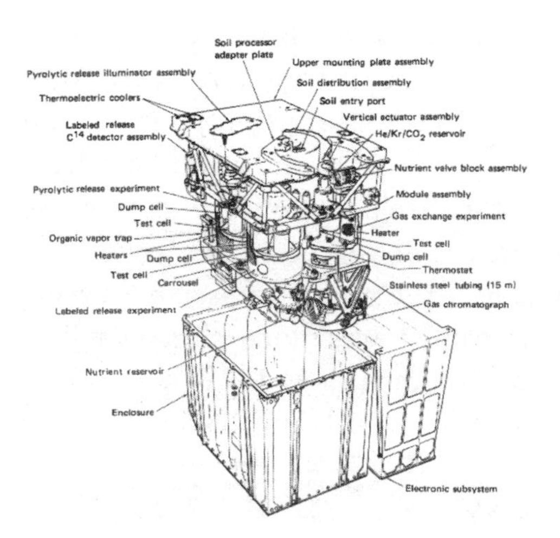

Figure 10.1. LOOKING FOR LIFE: The Viking life-science experiment was a small unit that incorporated four separate experiments, very advanced for the time, in a package weighing less than thirty-five pounds. Had the soil conditions been different, and there been something there to find, it probably would have worked. Some believe to this day that it did. *Courtesy of NASA/JPL.*

to make matters worse, there was one huge factor mission planners did not know: Mariner 9's pictures were not even as good as they thought.

When the dust storm that met that spacecraft settled, it was assumed that the images taken once the air cleared were the best that could be obtained by that camera. What the Viking planners could not know was that there was still an immense amount of dust in the air, softening every image they took. It was like having a silk stocking pulled over the orbiting camera lens, and the pictures were somewhat diffused. But to the folks on the ground, they looked just fine. Nobody would realize how degraded the images were until the Viking orbiters began shooting better pictures upon their arrival in 1976.

Also in play were the primary goals of the mission: to examine the soil on Mars for life-forms or prelife organic compounds. This goal affected landing-site selection more than anything except for concern about a safe landing. Would they be more likely to find life near an old water-carved feature? Near the poles? Mars receives intense solar radiation, despite its distance from the sun, and there was concern about the sterilizing effects of this on the soil. And the sampler arm, miracle of engineering that it was, could barely scrape a few inches into the soil, so it could not dig deep to find buried (and therefore possibly protected) microbes. It was a thorny problem, and it shows a bit of insight into how every element of such a mission has the potential to explode into a huge debate. This element did.

Leading up to the Viking landing-site decision, and after the demise of Mariner 9, the scientific community organized the "Planetary Patrol." Working with observatories like Lowell Observatory in Flagstaff, endless observations of Mars were made and reams of data examined in an effort to track cloud formation, dust storms, and anything else that could be gleaned from the limited abilities of Earth-based observing.

Viking planners were also hoping for images coming from the

Soviet Union's planned missions for 1973. Between orbital and, perhaps, surface photography, the NASA people might be able to gather some more data for Viking's cause. But these flights failed as had their forebears, and little data was returned—nothing at all of value to the Viking team.[1] The Americans would have to continue to stay the course alone.

So another element was thrown into the fray: radar. Using the largest radio dishes in the world, capped by the huge dish at Arecibo, Puerto Rico (a natural crater with a radio dish built in) they bounced radio waves off targeted areas on Mars. It was just one more way of identifying smooth (and, therefore, presumed safe) areas on Mars for landing-site consideration.

In the end, they would have to wait for images to come down from Viking's own orbiter cameras to make a final decision. This was cutting it close, but because the lander computers were somewhat reprogrammable, it was considered a worthwhile risk. Once the orbiters arrived at Mars in mid-1976, they immediately began sending back images of the surface below. And while they were spectacular, with resolution an order of magnitude better than Mariner 9's, there were a few surprises in store . . . not all pleasant ones.

One of the early landing-site candidates, a region called Chryse, had been selected after years of debate over the Mariner 9 images. It was an apparently flat, safe plain. But when the Viking images came through, Chryse revealed itself to be a rough, heavily incised riverbed. Filled with islets, craters, and channels, this was not the bedsheet-smooth area they had gambled on. To make matters worse, previous photographs had led photogeologists to theorize that Mars did not have many smaller craters. They were wrong, so very wrong. Much like the moon, it was filled with pockmarks of all sizes, from tiny to immense. Everywhere they looked, there were dozens, then, as one got closer, hundreds of them—any one of which was capable of wrecking the small landers. Hearts sank when the complexity of the terrain became evident.

In fact, this being the 1970s and with computers still an expensive luxury, a corps of graduate students over at Caltech (just a few miles from JPL) were pressed into service doing extensive "crater counts" from the Viking images in an effort to extrapolate what the nearby terrain might be like. It was an all-hands-on-deck effort.

Soon, after vigorous debate, the landing area was moved to a nearby region known as Chryse Planitia (Golden Plain). But once the photographic determination had been made, the ongoing radar survey showed that this region too was more hazardous than thought. The debate raged on until the last few days before landing, when a decision was reached. Chryse Planitia would be Viking 1's landing spot.

This left a decision to be made for the second lander. Site suggestions had ranged from the far side of the planet to the polar regions. And of course, once the macro decision was made for a region, the micro work began, attempting to assess the smaller dangers within. The area finally selected, through a similarly grueling process that lasted, again, until shortly before a commitment to landing, was Utopia Planitia (the Nowhere Plain), inside one of Mars's largest impact basins, and almost directly opposite Chryse Planitia on the other side of the planet.[2]

The torturous selection process for landing sites was almost alchemy; as much intuition and kismet as science and fact. And much the same can be said about the design of the life-science experiments, Viking's raison d'être, to which we now return.

One of the largest challenges facing the designers of the Viking life-science experiments was to determine what kind of life, what form of life, was relevant to seek out. How similar to terrestrial life-forms might any life on Mars be? This was the late 1960s/early 1970s, so the hardy life-forms that exist in places like the Atlantic seabed's hydrothermal vents had not yet been discovered and certain assumptions had to be made based on the state of the biosciences at the time.

The final design of the life-science experiments all depended on organic compounds in presumably benign Martian soil reacting to something. That something could be heat, light, and/or nutrients. And while from today's perspective the Viking package looks almost quaint, for its time it was an amazing piece of compact engineering, and the fact that it performed the tests it did successfully, regardless of the results, is astounding.

Each experiment within the suite had an individual container with associated heating elements and other apparatuses. Individual containers were fed soil from the sampler arm, which extended from the lander, scooped up some Martian dirt, then retracted and dumped the dirt into the experiment's container. It was all very high-tech for 1976. The experiments in question were:

The Gas Chromatograph/Mass Spectrometer: This device would sort through elements within a vapor given off when the soil was heated and then specifically identify them via molecular weight.

The Gas-Exchange experiment: This oven analyzed gases given off when a sample of Martian soil was "cooked" after having been fed a dose of chemical nutrients with water added. It looked for the hypothetically resulting metabolized gasses.

The Labeled-Release experiment: This device fed a small amount of nutrients to a soil sample, which were "tagged" with radioactivity, in this case carbon 14. The device looked for the release of radioactive CO_2 as a by-product of metabolization.

The Pyrolitic-Release experiment: This also utilized carbon 14 to measure possible photosynthesis. Light and water were added to the C14-laced atmosphere inside the experimental container. A period of theoretical growth was allowed, then, after the Martian "air" was evacuated from the chamber, the sample

> was burned to see if any of the C14 had been retained through photosynthesis.

Again, certain assumptions had to be made to move forward. One of them was the recipe for the liquid nutrients that were squirted into each of the experiments that needed them. The final solution used has often been described as a "chicken soup"–type fluid, the design of which was based on an Earth-based understanding of life. It was a noble effort. Other experiments used simple water.

Once the Viking 1 lander had succeeded in reaching the Martian surface, its first tasks were to create two images and conclude whether it was safe enough to proceed with its research program. Then the life-science experiments, weather observing operations, and the rest would have begun. But concerns over successful communication with Earth, about twenty light-minutes away, had resulted in the onboard computer, a for-the-time advanced 18-kilobyte (18k) unit, being programmed in such a way that if there were a failure to communicate with Earth, the lander could operate automatically. It would theoretically complete its primary mission without ground intervention for almost a month, sending data one way toward Earth as it proceeded. Fortunately for all concerned, the communication link among the lander, the orbiter, and Earth worked fine, and the probe did not have to be self-directed.

You might expect the first photograph from the lander to be of the dramatic, far-off Martian horizon. Indeed, many working on the mission would have agreed with you. But before the lander would image the horizon, JPL wanted to know what it was sitting on. After all, controllers had just experienced a three-hour-plus blind landing and were on the wrong end of a twenty-minute delay in receiving data. The first image would be not of the weathered Martian horizon, but of a footpad.

And what an image it was! As previewed in chapter 1, the

slow, strip-by-strip buildup of that first view of Mars was a suspenseful, thrilling moment. As the image was assembled on the giant video screen, one vertical strip at a time, it revealed pebbly, rocky soil, with a portion of the lander's footpad to the right.

The next image was a better press moment: a 300-degree black and white panoramic of the horizon. While Chryse Planitia may not have been the most daring landing site, it was still a dense tapestry of rocks, sand, and typically Martian elements, and would give geologists sufficient data to argue over for years. The next day, a color image was sent, and earthlings finally saw Mars for what it is: a fantastically alien environment with hauntingly familiar elements. We know that it is not home, but we can imagine it as such.

The early color images had problems, however; the sky appeared to be an earthly blue and the soil reddish-pink. Later, more highly corrected imagery shows a salmon sky and an orange-ish landscape. There is art as well as science in the discipline of planetary image manipulation, and it took those image-processing experts, pioneers in dealing with color images from space, a while to get it right.

Unfortunately, JPL was under extreme pressure from the press to get the color images out. What should have taken a week of color calibration was done in hours, and the sky hues were off. When the corrected version was later presented with a properly adjusted pink sky, the press corps booed. Some asked, tongue firmly planted in cheek, if the sky might be green tomorrow. It seemed that there was a strong desire to see Mars with a sky much like Earth's; not some alien tone of pink.

The scientists were dumbstruck. It had never occurred to anyone involved how important the images from Mars would be to people at large, nor the intensity of the emotional reaction to them. It was a PR lesson well learned that would pay off in later missions to the Red Planet.

During these early phases of the Viking lander's mission, a

meteorology boom had also been extended, and it began what would be over six years of Martian weather reports. The suite of instrumentation was designed to track the minutest qualities of the Martian day and night.

The values of the Martian atmosphere were confirmed: quite thin, at about 0.09 pounds per square inch (PSI) (compared to Earth's 14 PSI). This thin veil is composed mostly of carbon dioxide.

Lacking the weather-altering influence of oceans, or indeed any standing water, Martian weather turned out to be fairly predictable. The average temperature was -67°F, with a rare high of about +70°F and an equally rare low of -250°F (at the poles). Viking was able to measure much of this variation over a span of years, yielding very consistent results. At the lander's location in Chryse Planitia, the range was a comparatively moderate -22 to -139°F.

The thin Martian air changes temperature quickly when exposed to the sun, and with no oceans to break up the vast deserts, the resultant winds can be fierce. And these often become raging dust storms dwarfing any in the solar system, as witnessed by Mariner 9. Wind speeds measured by the Viking landers averaged about 10 mph and peaked at about 70 mph, though much higher speeds have been hypothesized for other parts of the planet. But with the very slight density of the atmosphere, the effects of even these high winds are much less than an equivalent speed on Earth—about a tenth as much.

Nonetheless, despite the low impact of these winds, it was a challenge when the Viking landers had to sit out a series of three global dust storms over their span of operations. The grit wrought havoc with the mechanical and optical systems. But both landers soldiered on, and the data they returned to Earth advanced our understanding of Mars one thousandfold.

Things with the Viking 1 lander were going swimmingly. Then the first gremlin took hold: the seismometer, designed to detect

"Marsquakes," was not working. The mechanism that had been intended to protect it during the violent assault of launch was still working, and too well. The detector was stuck in the safe position and was not responding to ground movement. As it turned out, a pyrotechnic device intended to disengage the retaining pin had not fired, and it remained in the launch configuration, which was useless on the ground. They could only hope that the Viking 2 lander did not repeat the trouble. It did not, and while it later provided for only one data point instead of the hoped-for two, the information gleaned from it was invaluable in evaluating the inner structure of the planet.

Then, on the third day, a problem that would be familiar to a later generation of Mars engineers cropped up: the lander began to think for itself—to rebel. The UHF transmitter, its sole link to Earth, was designed to operate in three power settings: one watt, ten watts, and thirty watts. For some inexplicable reason, it arbitrarily switched from thirty watts, its most powerful and effective setting, to a feeble one watt. The lander was getting bratty. The next morning, in keeping with the toddler analogy, it spontaneously switched back to thirty watts. After a few more tantrums, the transmitter stayed in the high-power mode until shortly before the end of the primary mission phase, when ground controllers reset it down to the ten-watt setting to conserve power. The ghost in the machine was silent . . . but only for a moment.

Soon the sampler arm, which was the only way to feed soil to the life-science experiments, became stuck. This ingenious device extended from just a few feet stowed to about ten feet extended.[3] It had been run out to nearly its full length to grab a sample, and then . . . it stopped. The worst nightmare of unmanned surface exploration of another world had just arrived. One of those little problems, an occurrence that could be solved by one single, swift kick to the lander, were someone there to do so, had taken hold.

The arm would not budge.

It didn't take long to assess the problem. A locking pin that was designed to fall out once the arm was deployed had not done so. Working with a twin of the Viking lander not far from the control room in Pasadena, technicians were able to duplicate the problem and come up with a solution. A few days later, all was ready. The command was uplinked; twenty minutes later, it arrived on Mars and was scheduled to be executed. The wait was excruciating. Eventually, images returned from the Viking cameras showed a pin-free sampler arm moving as designed. The mission went forward.

On July 28, eight days after arriving, Viking 1 reached out to grab some red soil. It was much slower than it sounds; these things are done with great caution and delicacy. The dirt was slowly winched back as the arm retracted and swung over the lander to deliver it to the experiments on board. But there was, of course, another problem. One of the experiments, the Gas Chromatograph/Mass Spectrometer, did not send a signal confirming delivery of a sample. On Earth, groans all around. There were a number of reasons it might have indicated failure: there might not have been enough soil to fill all the instruments, or it might be hung up in the feed trough, or the sample indicator might be faulty, or . . . or . . .

And there was a larger problem: each of these experiments was a one-shot deal. Once filled, they could not be emptied. So if they fired up the GCMS oven, and it was empty, they would have wasted one of the chambers on nothing but thin Martian air. The final decision was to try again, and dump some more soil into the chamber in question. That would solve the problem either way; it would fill for sure!

Then the sample arm jammed again. Oh, the thing had gotten some soil and begun to retract, but had stopped short of delivering the sample. Furthermore, it appeared to be unable to do any more work at all. This was not good.

After studying the problem on a ground-based twin of the lander, technicians saw that the arm, as it flattened into a metal ribbon, tended to kink-up under certain conditions. The incredibly ingenious design of the arm was its downfall. While the boom looked like a chrome pipe, it was actually constructed of a spring steel, not unlike a metal tape-measure. So, when retracted, the entire boom was wound onto a drum, flat like a ribbon. As it extended, it sprung back into shape as a tube and became rigid.

They went back to the control room. A new set of commands were issued, carefully designed (a) to give the thing plenty of time to operate, slowly, and (b) to operate only in a specified temperature range (to avoid extreme cold), and (c) to pace the commands in such a way that the motor was given a chance to operate in the most reliable fashion. To everyone's relief, it worked (and later in the mission the temperature restrictions were removed, without drama). Finally, the samples they had so coveted were delivered where they needed to go. It was time for Viking to fulfill its destiny: to determine if life, or at least organic compounds, existed on Mars.

Of course, this operation bumped up once again against the basic assumptions and philosophy of the mission designers. While everyone involved knew that looking in two fixed sites on a planet the size of Mars, grabbing random samples, and expecting to find something alive was a long shot, the press was not so shy. Expectations were high, and the pressure was immense.

With soil delivered, the devices were triggered. Nutrients were squirted, water added, ovens fired, and measurements taken; all with the greatest of care. What would the result be?

The soil samples were baking, the machines measuring. These samples had be selected with the greatest of care, as scientists would analyze images from the lander of surrounding terrain, pick an area, and carefully scoop up some soil. They even had the arm push a rock or two aside to expose "virgin" soil beneath, relatively unaffected by the sun and wind, and took a sample from there. It was painstaking work, and the science team was understandably anxious. Optimism on the part of these people ran the gamut from "I think we will find something" to "I feel it extremely unlikely . . ." (note the profound reserve inherent in the statements). But in their hearts, all those involved wanted the same thing: a strong indication of some kind of biological activity.

And then, faster than anyone anticipated, results were in. The signs of life were bubbling up inside the ovens and wetted sample containers. It seemed almost too good to be true!

There were, you will recall, three life-science experiments (the fourth such device, the Gas Chromatograph/Mass Spectrometer, was actually proficient at finding *any* organic materials, living or not). First, the Gas-Exchange experiment would indicate signs of living metabolism if microbes in the sample were flourishing inside its container and the enclosed environs. Second, the Labeled-Release experiment would measure decomposed organic waste if the microbes fed on the nutrient solution added to the sample. Third, the Pyrolitic-Release experiment would detect gases released from any synthesis of organic compounds in the soil.

And just like that, in that order, the dominoes fell. The Gas-Exchange experiment showed a buildup in pressure, a sign of activity within. Then the Labeled-Release experiment demonstrated a radioactive signal, seeming to indicate that something in the soil had metabolized and released the radioactively labeled gas. Finally, the Pyrolitic-Release experiment gave readings as well. The problem was that the readings from the experiments were not quite right. They ascended too quickly, and then decayed in an odd set of timings. Whatever was in the soil was responding all right, but not in the way predicted. It could be life, or . . .

After much head scratching, soul searching, and in-the-trenches analysis, a less appealing picture emerged. The final straw was that the gas chromatograph had not demonstrated anything organic in the release. It appeared that some kind of raw chemical reaction was taking place and *mimicking* life. All indications were that there was some kind of nasty oxidant in the soil (which was later confirmed to be a high level of perchlorate), which was reacting with elements of the experiments to provide false and misleading readings. Of course, not being there to look more closely, and to take samples into the lab and work them over with more sophisticated equipment, team members could only guess.

The team split into the "life" and "soil-chemistry" camps, with ill-defined lines between them. Some seemed certain; more straddled the divide. To this day there is a "we found life" camp, surrounded (and outnumbered) by a "we found chemistry in the soil" camp. The debate goes on, and will not be resolved until—possibly—the mission of the Mars Science Laboratory, now planned for a 2012 landing.

Then, one by one, the machines of Project Viking died. Working well beyond their predicted life spans, time and wear caught up with the spacecraft and they surrendered to Mars. First, the Viking 2 orbiter suffered a propellant leak and was deactivated by JPL controllers just twelve days shy of its two-year operational anniversary. Then the Viking 2 lander suffered a power failure and was unable to continue operations, ending its three-year, seven-month career. The Viking 1 orbiter met a more respectable demise: it lasted four years and two months on the job before depleting its maneuvering fuel. Then, unable to reorient itself to continue full operations, it was deactivated by JPL controllers.

But it is the Viking 1 lander's story that touches the heart. This plucky outpost was the last survivor of the quartet, and after almost six and a half years of operations, was at the time the grand elder of all things earthly on Mars. It had even been fondly renamed the Thomas Mutch Memorial Station, after a much-beloved member of the Viking team who had recently died in a mountain-climbing accident. But its long run ended in November 1982. Still sending back weather reports like a lone observer in a distant posting, it was due for a software update. With its plutonium power supply, it should have been good for many more years. But there was an error in the last batch of code sent by JPL; somewhere in the copious binary, there was an errant command that caused its radio dish to rotate down toward the sands below.

Like a loyal servant, it complied, and contact with Earth was lost. Despite diligent efforts from JPL, there was no further contact, and that was that. Nobody knows how long Viking 1 continued to "stare" into the cold desert wastes of Mars, awaiting another command that would never come. There may well be some electrical current flowing in its nuclear heart to this very day. . . .

Despite this unfortunate end, the science and discoveries of the Viking program would benefit future missions and fuel the next giant leap in Mars exploration: wheels.

CHAPTER 11

DR. NORMAN HOROWITZ

LOOKING FOR LIFE

The year was 1936. The first episode of *The Green Hornet* was heard on WXYZ radio in Detroit. The first radioactive element was produced synthetically. Adolf Hitler announced the first Volkswagen Beetle®. And Norman Horowitz arrived at Caltech in Pasadena, California. It was the beginning of an auspicious career at both Caltech and the Jet Propulsion Laboratory. He was a biologist by training, but his eyes was trained on the stars . . . and in particular, the planet Mars.

"By 1959, it was definite that the Jet Propulsion Laboratory was going to be a planetary science lab, and people began coming down [to Caltech] from JPL to see if there was any interest here in planetary exploration.

"I thought [life on another planet] was a plausible idea. Everything that was known about Mars at that time later turned out to be wrong, but [at the time] suggested that there was a good possibility of life on Mars. I had a choice of going into something . . . taking this golden opportunity to get involved in a new program. And that's what I did. It turned out to be very exciting. Of course, we didn't find life on Mars, but I'm glad I did it."[1]

Horowitz made a decision then and there that would affect not just his life, but the entire search for life on Mars. His move to JPL placed him in the Center for Planetary Exploration, where he would become one of the lead members of the Viking life-sciences team.

"The exploration of Mars became the key idea for a planetary program, for obvious reasons, and JPL set up a bio-sciences section to plan for the biological exploration of Mars, with an eventual lander. They asked me to come up and be chief of their section, which I did in 1965. There was a lot of work going on up there in trying to design instruments to fly to Mars for a biological search, and I got involved in that planning. Two of the instruments that eventually flew on Viking came out of that group. The Gas Chromatograph/Mass Spectrometer, which was probably the most important single instrument on the lander, was designed at JPL.

"When I went up there, that was already in process—it had been anticipated that this would be a useful instrument to have on Mars. What I did get involved with in connection with that instrument was making sure that there was a lot of ground-based experience with it. The instrument is based on empirical patterns of breakdown of organic compounds. You take an organic compound and you heat it until it pyrolizes—it breaks into smaller fragments due to the heating. These fragments can be identified by a combination of analytical steps called gas chromatography and then mass spectrometry. The only thing you have to identify the original compound you started with is the pattern of its breakdown products, and you try to infer the nature of the original compound from these breakdown products. There's not much general principle or general theory you can go on; you just have to have a library of results you can compare your actual results with. We did a lot of that during the years that I was there."

And this was the key to the search for life on Mars—trying to find a way to identify the building blocks of life by remote observation. To do this, Horowitz's team would have to build up a large database of similar reactions working here on Earth. It was not a trip to Mars, for which many of them would have gladly gambled their lives, but was the next best thing to going there.

"Another thing I did was to get the idea for the second biological instrument that JPL had on the Viking lander. NASA called it the pyrolitic release experiment; we used to call it the carbon assimilation experiment. It was an experiment that I developed with two collaborators, George Hobby and Jerry Hubbard. The point of this experiment was to carry out a biological test on Mars under actual Martian conditions. It's hard to convey in a few words the total commitment people had in those days to an Earth-like Mars. This was an inheritance from Percival Lowell. It's amazing: in pre-Sputnik 1 days, in fact, up till 1963, well into the space age, people were still confirming results that Lowell had obtained, totally erroneous results. It's simply bizarre!"

And that was the challenge. Horowitz knew by the time he moved to JPL that Mars was not Earth-like in ways that counted toward supporting life, but sometimes he felt that he had trouble getting others to understand it. Oh, they might pay lip service to the thin atmosphere, the extreme temperatures, and the voluminous solar radiation, but living deep in their hearts was a very different image of the Red Planet.

"A lot of people thought Venus was covered by an ocean. But that was speculative; in the case of Mars, they were making measurements and coming up with the wrong answers. Measurements were made on the 200-inch telescope by . . . a well-known astronomer—and they were completely wrong. This is just one example. And this was all based on the desire of people to believe that Mars was an Earth-like planet. It wasn't until 1963 that this began to unravel; the first step in the de-Lowellization of Mars occurred in 1963.

"[That] was one infrared photograph taken at Mount Wilson. It was an unusually excellent photograph, showing the infrared spectrum of Mars. It must have been a very dry night above Mount Wilson, a very calm night. They got this marvelous single plate, and it was interpreted by Lew Kaplan, who was at JPL, and Guido Munch, who was a professor of astronomy here . . . and Hyron Spinrad, a young postdoc working on Mount Wilson at the time. They showed, first of all, the total atmospheric pressure on Mars. . . ."

But even the coldest scientific data must meet with an emotionally charged challenge when presented to the broader community, and history had something to say about the subject: "Back around 1900 Lowell had estimated [Mars's atmospheric pressure to be] 85 millibars . . . so when the space program started, it was generally accepted that the surface pressure on Mars was 85 millibars, and that carbon dioxide was a small fraction of this; the rest of it was assumed to be mostly nitrogen, as on the Earth.

"So at least [life] was plausible. The Martian environment appeared to be Earth-like, but a very cold and dry Earth-like environment, an extreme form . . . with all the same elements, with water available and enough pressure so that liquid water could exist at least transiently on the surface. This was a difficult point, to get enough liquid water to support life. With 85 millibars, there was a possibility that you could have liquified water, at least for part of the day."

But the soon-to-be-infamous Mount Wilson data showed something very different.

"[Kaplan, Munch, and Spinrad's] findings showed that the surface pressure could not be 85 millibars. It looked more like 25 millibars to them. They also identified water vapor in the spectrum; that had never been seen before. They found very little water. And it was obvious that carbon dioxide was a big portion of the atmosphere and not a minor portion.

"Well, this turned out just to be the first step. The next big step came in 1965, when Mariner 4 flew by Mars and found that the surface pressure was more like 6 millibars! And that is the average pressure. And carbon dioxide is the principle gas in the atmosphere. Well, with 6 millibars, there's virtually no chance of having any liquid water."

By now the great Martian cities and canals and pumping stations of Lowell and others had died a quick and merciful death at the hands of Mariner 4. But there was still a chance for something smaller, more simple, and more realistic: "There was [still a possibility for life]. The main point up until Viking was water. And there were enough theoretical mechanisms for getting some water of the surface of Mars to maintain the remote possibility—although by the time we launched Viking, it was very remote—that there were either pools of brine or, after snow or frost there might be enough meltwater at sunrise to sustain a population of microorganisms. . . . [T]he real interest was in the possibility of having microbial life."

But, though the discoveries spoke loud enough for Horowitz and others like him to hear and adapt, it was a challenge to change the thought patterns of the broader scientific community.

"In spite of all these new discoveries, people were still building instruments to fly to Mars that were based on the terrestrial environment, and they were eventually approved by NASA. NASA was supporting these efforts. Around 1960, I got involved in one of them, one that actually later flew on Viking. We called it Gulliver at the time. It was invented by an engineer in Washington, named Gilbert Levin. It depended on an aqueous medium. Two other experiments that were being supported by NASA also involved aqueous solutions into which you would put the Martian soil and then use various ways of measuring the metabolism of the organisms. But after 1965, after the Mariner 4 flyby, it was obvious that the chance of liquid water on Mars was so remote that one had to plan for the contingency that there was no water—that if there was any life on Mars, it was living under conditions that were in no way terrestrial. So we designed an experiment that would work under Martian conditions and that involved no liquid water."

Old notions die hard, and the persistent idea of some kind of earthly life on Mars was no exception: "I think most of [this] was [because] people didn't want to give up the idea. And I agreed that, now that we had the capability, we would never solve the problem by just looking at Mars from the Earth. This was a classical problem, part of Western culture, the idea of life on Mars has been around for three hundred years. And here was the first time we had the ability to test it.

"Mariner 9 found an objective argument for flying to Mars, because [it] saw that Mars once had water on it. There are dry streambeds, obviously cut by water. All the geologists agree that they're water cut; there was water on Mars at one time. And you could say that, if there was water on Mars, then there may have been an origin of life, and that life may still be surviving. Now Mariner 9 was an orbiter . . . and up to that point, up to the time Mariner 9 took its photographs, I would have said the a priori probability of life on Mars was close to zero. It would have really been an irrational act to fly to Mars before 1971 to look for life."

But, as with his life-science experiments onboard Viking, studies on Earth would prove a valuable precursor to experiments on Mars. And few places on Earth were as close to the Martian environs as Antarctica.

"Another important thing I initiated at JPL [were] studies in the Antarctic. I never went to the Antarctic myself, but there was a microbiologist at JPL named Roy Cameron who studied microbial life of the world's deserts—he was traveling all the time. Just before I went up to JPL, I read a report of biological work that had been done in the Antarctic during the International Geophysical Year, around '58. . . . [T]here are areas called the dry valleys, actually ice-free areas. A team of microbiologists . . . got in there during the International Geophysical Year and they found that a lot of their soil samples were sterile; they couldn't find any bacteria. These dry areas are as Mars-like as you can find on the Earth. They're very cold and they're very dry. Roy brought back tons of soil. . . . [These samples were] used for a long time as standards during the testing of the Viking instruments."

But after all the things Horowitz has experienced and done, studying the most minute organisms on one world and seeking them on another, he still has a broad and revealing global perspective.

"I think that Mars exploration is quite important. If we are the only inhabited planet in the solar system, and there's only one form of life on Earth—I mean, when you look at the composition of living creatures and see that they all have the same genetic system and they all operate on DNA and proteins composed of the same amino acids with the same genetic code . . . then we're all related. The origin of life may have happened only once, and it happened here and no place else in the solar system. Or if it happened elsewhere, it didn't survive. I think this is a conclusion of really cosmic importance. If people become aware of this, then maybe they'll be less inclined to destroy the planet."

The exploration of Mars may indeed serve many functions. Let's hope this is one of them.

CHAPTER 12

RETURN TO MARS

MARS GLOBAL SURVEYOR

For twenty long years, Mars was left to slumber . . . six Soviet missions either failed outright or returned only partial results, NASA turned its attentions to the space shuttle and the outer planets, and nothing new from Earth landed on the Red Planet. The frigid surface remained untouched by human endeavor with only two Viking landers and a clutch of failed Russian spacecraft to mark the coming of the human race. Above the ruddy surface, a larger collection of machines remained, silent in their endless orbits.

Then, in September 1992, the United States reentered the fray with Mars Observer. Its goals were ambitious: survey the overall mineralogical and topographical nature of the planet, map the gravitational field, measure the magnetic field, and observe the atmosphere and dust within. It was a robust mission with great hope attached.

Originally intended for an Earth departure via the space shuttle, Mars Observer eventually left Cape Canaveral aboard a Titan III rocket.[1] Leaving Earth orbit, it headed off into an eleven-month cruise toward Mars. Just short of the first anniversary of the launch, in August 1993, the unthinkable happened: contact was lost with the spacecraft. Commands were sent repeatedly with the hope of reacquiring contact. Controllers waited tensely on the ground for any indication that the spacecraft had merely wandered off course or turned off axis and would recover. But it was not to be, and silence remained the sole result.

Hands were wrung and heads hung low at JPL. Surely after all these years, with the last Mariner failure occurring decades earlier, technology had progressed to the point that spacecraft en route to other planets should succeed, especially those headed to this close neighbor of Earth. But alas, this was not the case. As far as is known, part of the propulsion system failed, disabling the probe. Whatever the case, Mars Observer ceased to observe, and went silent forever.

It was noted in the postfailure investigation that the spacecraft had been converted from an Earth-orbital satellite. Some of the systems (propulsion specifically) might not have been up to the rigors of interplanetary travel. Cost-cutting measures had caught up with optimistic planning, and the entire mission was now a write-off, save for some data acquired along the way. It would not be the last time that misguided frugality bedeviled JPL.

But this was not the end of the road for orbital observation of Mars in the 1990s. In 1996, a Delta rocket roared out of the cape with a 2,200-pound payload, almost exactly the same mass as the ill-fated Mars Observer. It was Mars Global Surveyor (MGS), which would reclaim the mantle of Mars exploration for America. Built by Lockheed Martin, the MGS craft was a simplified version of the Mars Observer. A new high-resolution camera was onboard, along with a suite of other instruments that would replicate much of the capability lost when Mars Observer died.

Once en route, and when the spacecraft deployed its solar panels, there was one mishap discovered: apparently, during the brutal stresses of launch, or upon opening the solar panels, a small damping strut (used to regulate the swinging-open of the solar panel, much like a screen-door closer) snapped, and the stray part had lodged in the fuselage and prevented the panel on one side from fully opening and locking. This was a problem, as the solar panels, once deployed and configured into a V-shaped pattern, were critical to the aerobraking maneuvers once the ship reached Mars. Ground teams worked overtime to come up with a solution.

It took almost a year of drifting through the Great Dark to arrive at Mars, but MGS continued without additional mishap. Perhaps things were looking up. As the craft entered Martian orbit, commands were uploaded to change the course of the probe to enable aerobraking. This relatively new technique would deliberately force the ship into the upper reaches of the Martian atmosphere and slow it down and alter its altitude, over time, from over 33,000 miles to only about 280. MGS would be the first spacecraft to try this risky but elegant technique at Mars.

The reasons for aerobraking were simple. The cost-effective Delta rocket did not have enough thrust to loft a full fuel load for a more traditional trajectory and rocket-braking maneuver to reach Mars. In the case of MGS, the trip involved a long loop around the far side of the sun to get to its target, but it would result in a lower velocity once it arrived. Therefore, less fuel would be needed to slow it into a lopsided orbit, and aerobraking by skimming the atmosphere would trim-up the arrival into a proper orbit to achieve its goals.

Prior to this, JPL personnel had to come up with a solution to the broken strut. The solar panels were actually a part of an aerodynamic design for the craft, interacting with the wispy upper atmosphere to help slow and lower its orbit. With one of these panels hung up, not only could the spacecraft be unstable (even in the tenuous upper atmosphere), but the pressure of ongoing aerobraking maneuvers could further damage the panel mount, and maybe destroy the entire probe. The solution? Rotate the panel 180 degrees to present the solar-power-generating side to the winds of the upper atmosphere. Not only would this avoid further damage, but it might also act to help the panel latch into a locked position.

Aerobraking is a slow process, and it took nearly one and a half years to accomplish. The Martian atmosphere is thin at the surface, and exponentially more so at high altitude. But to avoid damage to the craft, only the fringes of the atmosphere must be allowed to drag on the ship. *Caution* was the watchword for this portion of the journey, as aerobraking was critical to success, and without careful completion of the maneuver, the mission would fail. This would be a kinder, gentler aerobraking approach than originally specified, in hopes of suspending further damage to the delicate craft. It worked.

Patience is a virtue often rewarded in space exploration.

Finally, in March 1999, the desired orbit of about 280 miles was reached. At this altitude, MGS would circle the planet every two hours. The orbit was polar in orientation, that is, moving from north to south instead of along the equator. While more challenging to accomplish, the scientific yield would be much higher, as with this orbital path, every part of the planet would repeatedly pass underneath the cameras and other instruments.

Soon the mapping runs began, with high-resolution images flowing in hourly. The scientists were ecstatic. The onboard cameras, a new high-water mark in camera design for a Martian probe, showed objects as small as eighteen inches across. At centers around the country, working in tandem with JPL, breathless researchers eyed each new photo pass with glee. While the images from the Viking orbiters had been striking, these were exponentially more detailed. Additionally, due to the polar orbit, MGS eventually covered the entire surface of the planet in approximately the same lighting conditions on each pass. The results were stunning.

While it had been clear that wind, sand, and water (in some form) were at work on the Martian surface since Mariner 9, these new pictures allowed planetary geologists to refine their theories about weathering, hydrology, and atmospherics on Mars.

Early interpretations of the imagery showed more detail of the landforms that had so baffled scientists, confirming that these were in fact wind- and water-sculpted formations. This was exciting news, for it indicated not only an active weather system, but also evidence of vast amounts of water somewhere in Mars's past sufficient to carve out huge masses of soil and rock. Until then, orbital data had not formed a clear picture of what might have been at work earlier in the history of the planet. But here it was—in stunning detail—evidence of huge masses of water sometime long ago. And where there was water, there could have been—and might still be—life.

Other instruments onboard included a sophisticated laser altimeter, allowing MGS to measure the elevations of Martian topography accurate to *one foot*. This allowed planetary scientists to not only map the rocks and sand of Mars, but also re-measure areas of interest across many years, sometimes catching differences in height that indicated erosion and soil movement.

A thermal spectrometer allowed researchers to see the planet in infrared, which demonstrated yet more evidence of large masses of water in the past by revealing topographic evidence of ancient hydrothermal activity and water flow. It further indicated large deposits of hematite, which often originates in large bodies of standing water.[2]

A magnetometer measured Mars's weak magnetic field, which, unlike Earth and Mercury, does not originate from a heavy, central, iron-rich core. Rather, the magnetic masses are concentrated in various areas around Mars, indicating massive volcanic activity early in the planet's history. Further data showed a deeply layered crust on Mars, reaching to a depth of over six miles. This indicated the likelihood of a smaller molten core than Earth's.

Although the two hemispheres of the planet appear to be very different—the top half is smoother and lower in elevation, while the bottom half is much more intensely cratered—it was now apparent that there were plenty of craters in the northern areas as well, but many had been buried. But buried how? Making things more complex, the vast majority of the surface was underlain by volcanic rock. So it was, at one time, a highly active planet in geological terms. This confirmed widespread volcanism, not just in the region of the giant volcanoes evident to the north.

Also, while not scientifically significant in the traditional sense, MGS photographed the Cydonia area of Mars, which a Viking orbiter had imaged in 1978. At that time, the first pass by Viking showed an area that vaguely resembled a human face. While the planetary science community was unmoved, it created a popular sensation, championed by some less-than-stellar pseudoscientific personalities. And despite the fact that later Viking images of the area seemed far less facelike, a myth was born. Some wanted desperately to believe that it was an artificially created structure. Then, in 2001, MGS imaged the area again. It was a spectacular shot, but not appealing to the true believers. The region, while still eerie in appearance, was clearly the home of a large erosional feature—a result of weathering, not intelligent design. It no longer strongly resembled a face; any likeness was vague at best.[3]

Finally, in one of its last acts of great scientific return, MGS produced images that seemed to indicate recent water activity. In December 2006, gullies with fresh sedimentation were spotted

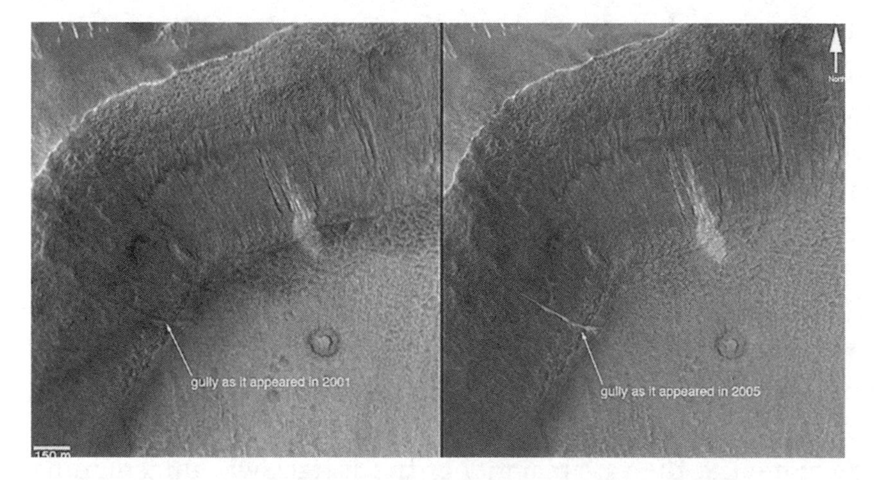

Figure 10.1. WATER HERE AND NOW: Seen here are two images from the Terra Sirenum region. One was taken in 2001, the other five years later in 2006. The arrow points to a feature that appeared in between the taking of the two images. It was as close to a "smoking gun" as ever happens in planetary exploration, a probable sign of water. *Courtesy of NASA/JPL.*

inside two craters, Terra Sirenum and Centauri Montes, and this would have to have been caused by water flowing within the last few years, arguably sometime between 1999 and 2001. This was staggering news, as it had been generally thought that whatever water had been on Mars was long gone, or permanently frozen deep underneath the crust. Some mechanism must have heated and released the water required to accomplish this. Whatever the case, this was a major discovery.

As with most of the spacecraft launched from JPL that reach their destination successfully, MGS was not yet finished at the end of its primary mission. The machine was well designed, well built, and well handled, and had much more to offer. The mission was extended three times past the planned 2001 end date, and MGS returned data until November 2006. Then, it abruptly stopped speaking to ground control.

This failure occurred after a series of commands had been sent to the spacecraft to reorient its solar panels. The onboard computer signaled a series of alarms, including some related to its orientation, but then reported that it had stabilized. That was the last message sent earthward by the probe. Various attempts were made to reacquire contact with the spacecraft, and three days after it went silent, a faint signal was received indicating that MGS had gone into "safe mode," a computer condition that occurs when the situation aboard the probe is not as expected. Nonessential systems had been shut down and the craft was awaiting additional commands.

JPL does not give up on its interplanetary emissaries easily. Controllers even pulled a later arrival to Martian orbit, the Mars Reconnaissance Orbiter, into the rescue effort, attempting to snap a picture of MGS to observe what condition and orientation it might be in, but this was not successful. MGS's younger sibling would not be of aid.

In the final analysis, it appears that a programming error—a human failing—caused the ship to turn into an improper orientation with regard to the sun, overheating the onboard batteries and causing them to fail. With no power (as the solar panels were not correctly aimed at the sun) and dying batteries, the craft had little time to reacquire a signal with JPL. And while its programming likely resulted in its searching for a proper orientation to talk to its masters, this software did not include a way to keep itself safe from the rays of the sun, destructively hot even this far away.[4]

While the craft died what was ultimately deemed an unnecessary death, the mission was a vast success. MGS operated four times beyond its primary mission, and for almost ten years, longer than any other Mars orbiter. It provided evidence of huge masses of flowing water on Mars (heretofore thought unlikely, despite the visual evidence), and found deposits of minerals that further indicated massive water flows in the past, helping to select future Mars rover landing sites.

The oversights that resulted in its demise also triggered a thorough review of the safety features loaded onboard future spacecraft, and this is how places such as JPL refine their approach to space exploration. While all contingencies are considered as thoroughly as possible in advance, it is by failures that we learn to avoid such future debacles. And this paved the way for even more successes ahead.

CHAPTER 13

ROBERT BROOKS

IT TAKES A TEAM, MARS GLOBAL SURVEYOR

Bob Brooks is an icon. He wears his trademark dark T-shirt and flannel overshirt, with his long hair tied back in a ponytail. The hair is not a conceit, nor is it an homage to the 1960s. He says that it's just easier to see what you are doing that way.

When not at JPL guiding spacecraft to their fates, he can often be found in a darkened room in his Van Nuys home, fiddling with high-powered lasers and other dark technical arts. He is an inveterate tinkerer, a thirty-three-year veteran of the lab, and a passionate voice for planetary exploration.

"I went to USC, got my bachelor's in astronomy, and then a friend arranged for me to get an interview with his boss at JPL. . . . [N]ext thing I knew, I had a job working on Voyager. I stayed on that mission for several years. It was really fascinating, traveling to the outer planets. Then there was a new project called Mars Observer, and I got very interested in that. It was the beginning of a long-term involvement with Mars.

"Well, we lost MO. There was a lot of concern about that. So the lab wanted to try to do a better job next time, and the director at NASA headquarters sold [NASA] on the idea that we would actually have two projects: one would be Mars Global Surveyor, which would be basically to do the same things that MO was supposed to do, and the other was something called MSOP, or Mars Surveyor Operations Project, which was not a flight project, but a way to start development of a multi-mission operations center at the laboratory, which would result in cost savings and higher efficiency.

"So that's what we did, we built MGS up, and in the same time frame we also built MSOP, in order to operate it. As we built MSOP, more and more projects afterward got interested in using the things that we offered, not just the software, but the people. So at one point MSOP was flying several projects at once, and doing it very inexpensively. It was a great way to manage spaceflight.

"I was one of the founders of MSOP, and was asked to come up with a good operational strategy that would save money and work in a multi-mission environment. Working with two colleagues from Mars Observer, we had developed an entity called the ASP, or automated sequence processor, which is still in use almost twenty years later. This was a way for remote science and spacecraft team users to send commands in from wherever they were, anywhere in the world. These commands would be automatically processed through the ground system, and if they were good, then they would be prepped for radiation to the spacecraft. On the other hand, if there were problems with them, then the ASP software would send back a nasty-gram to the requester saying 'You have to fix this!' This was something that was available 24/7 and required no personnel sitting and validating commands, it just ran all by itself. That was a big deal at the time, and everybody loved it. It saved anywhere from 1.5 to 5 million bucks per project, per year. Ultimately, we could have four or five projects running at once. Previously this would have been prohibitively expensive, but this allowed the projects to come in and very efficiently run their operations."[1]

MSOP was a revolution in unmanned spacecraft command and control. But "JPL'ers" live to explore, and that means flying specific missions. And Mars was calling across the void to Brooks: "But with the MSOP operating well, I wanted to go back to Mars and help to redeem ourselves from what happened with MO. So the MGS mission was that chance. One of the unique things that we decided to do with it was called *aerobraking*, which was very new. We had done a short dip into aerobraking at Venus with Magellan, but for MGS we had to conduct this risky maneuver for months to skinny-down the capture orbit, which was forty-four hours long, to our two-hour mapping orbit.

"Now, aerobraking is a very complex and, potentially, personnel-intense procedure. But this automated sequence processor that we had designed made it possible to do this cost effectively. So while the aerobraking took several months, we picked one person who was the best at it, and he built up an input file for the ASP software which generally automated the procedure. Once we made sure that it was okay, the commands could come through in the middle of the night, the ASP would grab it, run it, finish it, get it ready to go, and send it on. It sounds complicated, but it actually saved a lot of steps, a lot of work-hours, and allowed us to pull off missions that would otherwise have been too expensive."

Of course, MGS had one major issue during aerobraking: the broken solar-power-panel support.

"What happened was the yoke, or frame, that connected the solar panel to the spacecraft body broke; it snapped on one side, and so if we did aerobraking the way we originally planned, we would have had problems. Remember that aerobraking involves flying down into the atmosphere a little bit and using the drag to slow down the orbit, and slowly your orbit gets smaller and smaller.

"Now these orbital spacecraft have solar panels spread out to the sides. No avoiding that, they need the power supplied by the sun. So the plan was to actually configure the solar panels facing in a certain direction, so when we dipped down into the atmosphere, the force exerted by the atmosphere actually pressed evenly against those panels, using them a bit like resisting wings. But with that broken mount, the force would have snapped the solar panel off, and that would have been the end of that spacecraft! So instead we had to prepare the spacecraft for each dip into the atmosphere by rotating the broken panel 180 degrees, so the force was pushing now against the break back into position.

"As you are doing this you're worried about lots of things, including the motors that turn the solar panel. These things won't work forever. And because you don't know when you're over-amping (and potentially damaging) something, you want to minimize that kind of activity on your spacecraft so you don't just use up all of your resources before completing your assigned mission.

"Nonetheless, during each aerobraking maneuver, and this took months, we had to rotate the broken panel away from the sun, dip into the atmosphere, back it out, and rotate the panel back around. Ultimately we were fine, it was just a matter of coordinating all this activity as you did this aerobraking stuff. In the end it was actually quite the education for everyone, and that was really how aerobraking got perfected and it's something that's been used many times since. But as the first, and with that mechanical problem, MGS was what forced us to do some really crazy things to make it happen."

Once MGS got into its stable, round orbit, the images—and discoveries—started to pour in. Many were memorable, even these many years later: "I remember a picture that showed a crater wall with what looked like a dried streambed from liquid water that appeared to have gushed out of the wall. That was just one thing. We found all kinds of neat stuff using the science instruments on that spacecraft. But in the end, MGS was really about efficient planetary operations, and therefore low cost, and about trying to recover the science that we lost because of the failure of Mars Observer."

But the thrill of discovery was punctuated with some true white-knuckle moments . . . ones that required immediate attention: "So I'm in bed one night, this was during aerobraking, we were in a forty-hour orbit, and had a long way to go. The phone rings at about 2 a.m., and I picked up and it's Glenn, the project manager, and he's asking if I have any idea why the spacecraft just executed the same course correction maneuver twice and we're headed for an impact on Mars.

"That's an attention-getter, and I woke up really fast. . . . I said, 'No I don't have any clue, but I'm on my way in right now.' I jumped out of bed, got dressed, and hopped into the car. Soon I'm doing ninety miles an hour from my house to the lab, which is twenty-three miles. I think I got there in about fifteen minutes! I walk in and Glenn is standing there in the conference room looking at some telemetry, asking what I thought was going on. I said I had no idea (I had just woken up, after all), but I'm sure that it's some part of my ground system, so I just stayed there to dig into the problem. We had several hours, but it was still necessary to fix the problem (we were headed for impacting Mars), and so he woke up people to do an emergency maneuver."

"Long story short, the spacecraft was saved because the flight team got up, everybody came in, they figured out what they had to do to get the spacecraft back into a normal orbit, and everything was fine. As it turned out, we discovered that a remote software delivery had installed a new memory map for the spacecraft, which tells it where to store things in the spacecraft's computer. This resulted in the maneuver commands being put into two different locations in the spacecraft's memory and those commands being executed twice . . . and that's not good."

"Once I figured out what happened, the solution to that was really quite simple: don't let folks just deliver software when they feel like it. This was all a matter of timing. The error occurred because the ground software had been changed at the same time as a sequence was being processed through it, changing the intended result. I know some people were very unhappy about my policy changes because they were so used to doing things loosey-goosey, but it became a strong, hard-and-fast rule. 'You will coordinate all software delivery with the operations team so that this doesn't happen again.' If this had happened during one of the shorter [later] orbits, we would have lost that spacecraft."

But, thanks to personal dedication by the flight team (and lax speed-limit enforcement of some Los Angeles–area freeways in the wee hours of the morning), MGS lived to fly on into history. One of its major discoveries was the verification of a lack of a global magnetic field around Mars.

"Either Mars doesn't currently have a molten core, or if it does, it's not a gigantic thing like the Earth's. If it exists, then it's probably heated by friction but it is too small to cause convection. And without this, you won't have a global magnetic field. Note that at one time Mars had to have had a fairly substantial molten interior, because you have these huge volcanoes on it, at least four of them. But for now we don't see much of a magnetic field and what we do see is localized and regional."

And then, in November 2006, it was suddenly over. Unplanned, unanticipated, the mission came to a sudden halt: "The mission ended, but it ran four extended missions and operated for about seven years, so it was not like MGS died when it was at its prime. It was an old spacecraft, and while it was still producing perfectly good data, it had paid for itself several times over. Recall that each of those missions lasted a Martian year, or about two Earth years, so we got our money's worth out of this; it's all gravy after the primary mission, certainly after two or three extensions."

That said, Brooks perceives a sacred trust within the JPL community, a bond between the taxpayer and those who fly the unmanned missions of discovery: "We have a mantra, and that is that spacecraft don't belong to us. It's not my spacecraft, it's not even really NASA's; it belongs to the American people, and because of that, we have to take very good care of it, and we have to get out of it everything we possibly can. We must basically milk the cow until it's dry, and that's exactly what we do with all of our missions. Look at what we did with the Mars Exploration Rovers. Eight years in, one of the rovers is dead, but the other one is still going along and doing its thing. That mission was supposed to finish years ago, and yet still it's going. But let's go back even further: look at [the Voyager mission, both Voyager 1 and Voyager 2 are] still running, for probably another twenty years, until the RTGs' [Radioisotopic Thermal Generators'] power supplies aren't working anymore. That was launched in 1977, and it's still sending back data a quarter century later! That's value.

"When I go out and do talks, one of the things I tell people is that the cost of a large mission is, say, 750 million bucks. So you look at the population of the United States, and it's about 300 million people, and so for a bit over $2.50 per head, not even the cost of a Big Mac®, you pay for your part of a fantastic mission of discovery. Just skip one burger for a day in twenty or thirty years and you've done your part to pay for the mission. I think it's worth it."

And while McDonald's may not enjoy the same sentiments, it seems to be more than a fair trade. I'll give up a burger for *two* days myself.

CHAPTER 14

ROVING MARS

SOJOURNER, THE PATHFINDER

The rock smelled. Of course, it was supposed to smell. As she took repeated whiffs of it, there was little to compare to. Sojourner knew only *this* rock . . . it was the first one she had smelled since arriving on Mars just under three days ago.

It didn't smell in the traditional sense that humans or animals would experience. Sojourner's "nose" was a highly sophisticated device, a marvel of robust yet lightweight and compact engineering called an Alpha Proton X-ray Spectrometer (APXS). This little machine, just a fraction of Sojourner's twenty-three-pound overall weight, was designed to analyze the chemical composition of Martian rocks using protons it emitted to excite the elements making up the rock in question. Using this, Sojourner could smell just five things: sodium, magnesium, silicon, aluminum, and sulfur. But her designers had decided that this would be enough, and it was. For, while it could not distinguish between a Bolognese sauce and raw onions, it would be able to identify the basic makeup of a rock. And that was far more valuable on the surface of Mars than an epicurean snout.

It was not a fast process, however. Sniffing this first rock, affectionately named "Barnacle Bill" by JPL scientists after its surface, which appeared to be encrusted with the fishy crustaceans, took over ten hours to complete. But Sojourner was patient; she had already endured eight months in the cold darkness of space and a rough-and-tumble arrival on Mars. Ten hours at one rock—

an interesting one to boot—was little in the grand scheme of things. So she took her time.

Just under three days, or *sols*, previous (a sol is a Martian day, just a bit longer than an Earth day at 24 hours and 39.5 minutes), Sojourner had arrived on Mars in dramatic fashion. After her long cruise through the interplanetary void, she had skipped entering Mars orbit as had her predecessors, Vikings 1 and 2, and shot straight into the Martian atmosphere. It was all part of NASA's "faster, better, cheaper" approach to unmanned space exploration, which was being practiced at the time. Officially called the Discovery Program, it was an attempt to speed up these projects with smaller, lighter, less expensive, and more focused scenarios. And it worked—at least on this mission.

She entered the thin air above Mars safely affixed to the Pathfinder lander, later known as the Dr. Carl Sagan Memorial Station. The lander, in turn, was enclosed inside a protective aeroshell with a heat shield at its back. As it plummeted toward the red soil below, a parachute, with large areas removed to allow the supersonic Martian air to stream through without damaging it, was deployed to slow the craft to a survivable speed after about two minutes. Shortly thereafter, the heat shield was kicked away to expose the lander, which was winched down from the supporting structure on a cable to dangle sixty-five feet below.

At about a minute before touchdown, and one mile up, a radar altimeter was turned on to track the distance between the lander and the hard ground below, now only about thirty seconds away. Then, with less than ten seconds to go, a cocoon of airbags, looking like a diseased cluster of large gray beach balls, was inflated, completely surrounding the lander. This was so unlike the more traditional Viking entry and approach profile it surely raised eyebrows (and perhaps blood pressures) throughout the space community right up until this point. A few seconds later, small rocket engines ignited, burning for only two seconds, drastically and immediately slowing the small craft further. Then the cable connecting it to the parachutes and engines was cut (this allowed the parachute to drift away and prevented its snarling with the lander), and the craft dropped, free of any influence save for Martian gravity, to impact the surface at about forty miles per hour (far faster than previous surviving craft had). The beach balls absorbed the force, and the entire assembly bounced back into the sky about forty-five feet . . . then a bit less, and a bit less . . . In the end, it took fifteen bounces, rolling along at about 30 mph, for Pathfinder to come to its final resting place.

And there it sat, slowly rolling to one flat side of the airbag assembly. Once it settled, the airbags deflated, aided by a triggering device that unzipped the insides of each bag cluster. Once the beach balls were merely flattened, deflated memories, winches built into the lander retracted them to clear the nearby soil and haul them underneath the lander. Mars Pathfinder was a spacecraft that cleaned up after itself.

After nearly an hour and a half, the lander unfolded. It had a base with three "petals," each a solar panel, and one with departure ramps for the small rover within. The action of lowering the petals finished the job of forcing the lander to sit upright. It was still dark, as the machine had descended to the surface early in the Martian morning—about 3 a.m. local time. Not much more could be done until dawn, so everyone had to be patient. It was July 4, 1997. Millions of miles sunward, a good part of one of Earth's continents celebrated with explosives and rockets similar to those that had delivered Pathfinder to the arid plains of Mars. But here, in the –100 degree wastes, quiet reigned in Ares Vallis.

Ares Vallis had been selected from a long list of candidate landing sites. Years of analysis of the hundreds of thousands of images sent back from previous Mars probes, especially the two Viking orbiters that had arrived in 1976, had given planners almost too much information to sift through. The problem, though, was less with bulk than with detail. The resolution of the Viking cameras, while excellent for its time, was too low to give a really good idea of what lay below. Anything smaller than about twenty-five feet across was a mere speck. So, while these pictures gave a good general idea of what kind of landscape inhabited broad areas, they were not of sufficient quality to select with exactitude.

But the planetary scientists were undeterred. A lot can be inferred by what *surrounds* an area. The two Viking landers, designed and built in the computer-challenged 1960s and early 1970s, had landed in areas that were barely known (the photos of Mariner 9 and a handful of Viking orbiter pictures were the best they had then), and safety was a much higher priority than fascinating geology and topography. These machines, sometimes referred to today as "Big, Dumb Landers" had to set down while flying blind with minimal feedback from the Martian ground and none from Earth.

Technology had come a long way since then, and even off-the-shelf components (which was what composed most of Pathfinder's data capabilities, another budget-conscious choice) were a quantum leap over what had gone before. So with this added assurance, and the reasonably good images from the Viking orbiters, the Pathfinder team had made its choice of where to land.

Ares Vallis was a flood plain, adjacent to a wide channel, that appeared to have been cut by flowing water—lots of it. The exact landing area was a spot where this channel approached a delta, and eventually opened into Chryse Planitia, the area where Viking 1 had landed (though still over 525 miles distant). If this was what it appeared to be (and it was), lots of different kinds of rocks and minerals should have been washed out of the channel and into the delta. The hope was that Pathfinder would have lots of interesting samples to choose from.

Once the sun rose over Ares Vallis, temperatures shot up to a comparatively balmy 10°F. It was now light enough, and warm enough, to begin operations on the lander. The first images were obtained and sent back to Earth, a twenty-minute trip. The meteorology package was also activated, and weather reports began streaming in. A quick analysis of the images showed that one of the airbags was not fully retracted and might impede the rover in its operations. This was quickly remedied, and Pathfinder prepared to take more pictures and spend its first full night on a frigid Mars.

There was a heart-stopper at 10:30 that night when the computer onboard the lander stopped sending information. Then, at about 3:20 a.m. the next day, signals received seemed to indicate that for some reason the computer had reset, or rebooted, itself. The reason was not immediately apparent, but it was working again. Everyone breathed a sigh of relief. It was the first, but would not be the last, technical cliff-hanger of the mission.

On sol 2, Sojourner was ready to roll. The command was given: the tiny rover, only about twenty-three pounds in operational trim, "stood up" and rolled down its ramp. Ever so slowly it began to move. Sojourner's optimum speed was about one half inch per second, but it would navigate the ramp down to the surface much more slowly than that. By the end of the day, it had worked its way to the bottom of the ramp and waited patiently for orders. They came at dusk: sniff the soil right where you are. Sojourner spent the night doing just that.

The next day, at 3:45 a.m., JPL controllers "woke" Pathfinder with the song "Final Frontier" from the TV show *Mad about You*. It seemed fitting. Instructions were uplinked to Sojourner to prepare it for an experiment in "soil mechanics," a fancy way of asking, "How does Martian dirt work?" So the rover was instructed to lock five of its six wheels and turn the sixth one, first one direction, then the other, grinding away at the dirt. By observing the reactions of the surface to this abrasion from the tiny stainless-steel wheel, the geology team on Earth would be able to divine much about the real estate close by the Pathfinder lander.

At the conclusion of this Martian version of smoking tires, Sojourner would begin its historic traverse to Barnacle Bill, the first of many rocks to be visited in the immediate area. Now, the word *traverse* might be misleading . . . it conjures thoughts of a long overland voyage. In this case, the total trip from Sojourner's position at the base of the ramp and Barnacle Bill was a slim fifteen inches! Still, this was a new technology on a new mission on a distant planet, so everything had to be planned with care.

Sojourner started her short journey. While this was under way, the lander began work on a so-called monster-pan, a complete 360-degree sweep of the landing area. The results were spectacular. After years of studying the Viking lander images from the 1970s, these new, highly saturated high-resolution pictures were simply breathtaking. They showed a whole new Mars. Not only was the topography and weathering different from that shown in the Viking landing zones, but the details of rock surfaces, types, and general weathering were far more evident.

The images in general were so superior that when combined with orbital photos from the various probes that had been investigating Mars from high above, assumptions could be rapidly made about the surrounding terrain.

One thing became clear right away: their interpretation of the landing area as resulting from water flows was spot-on. As one scientist put it, the flooding was so catastrophic that it could have filled Earth's Mediterranean basin. That's a lot of water for a desert planet. Water-transported deposits were seen nearby. Rocks such as Barnacle Bill had small "moats," or eroded depressions, around them. Areas of bright and dark soil were seen, indicating wear. And the rocks nearby appeared to be of differing origins, if not differing types altogether. It was, as one scientist phrased it, a geological grab bag. And this offered opportunities to have Sojourner "sniff" a variety of nearby rocks, presumably washed down from different regions in Mars's wet and wild past, to build a better picture of the planet's history.

As if to underscore this, the next rock visited, Yogi, was markedly different from Barnacle Bill. It was more primitive, not having experienced the heating, cooling, and general geological nightmares that the previous rock had.

These were exciting times. The first machine to land on Mars in twenty years was operating perfectly, and gathering reams of new and detailed data. And as usual, it was exceeding expectations.

By the time Pathfinder had been on Mars for six weeks, the mission was still proceeding well, but some glitches had popped up. On August 16, the flight computer on the lander had reset itself. It had not asked "DO YOU MIND IF I REBOOT?" or sent any other indication; it simply restarted of its own accord. Why this happened was not known, but at the time it was suspected that the temperature extremes experienced on the planet this time of year were severe enough to cause issues with the computer's circuitry. In any case, JPL worked the problem, sent a command to the lander to ensure proper aim of its high-gain antenna, and through this process reestablished contact at about 10 p.m. on the seventeenth. The sense of relief was palpable. However, this was neither the first nor last time this problem would rear its ugly little silicon head, and each time it did, the concern increased.

Soon controllers were receiving images from the lander again, and immediately another problem became apparent: Sojourner, still not far from the lander, was stopped on a rock called Wedge, and for obvious reasons. It had been heading toward another rock named Shark, and the onboard sensors, ever vigilant against hazards, had sensed a tilt that was more than it was willing to accept. The small computer shut down the rover and waited patiently for advice from Earth. After an intense conference, the team worked out a new course and sent the commands skyward. Sojourner headed off to Shark and the area called Rock Garden beyond.

Sojourner was traveling moderate distances unassisted now, as had been planned. Much of the Pathfinder mission was a test bed for future missions, so it was important to learn as much as possible while the rover was operational and within view of the lander's cameras. There were glitches of course—false stops, occasional digital confusion, and misaimed trajectories. But overall, Sojourner was proving to be a tough, smart, and plucky little rover.[1]

By the end of August, ice clouds were seen in the surrounding skies, and the sunsets were picking up color. Blue sky was observed around the sun at some of these times; this is due to the Martian dust scattering the blue wavelengths. Temperatures were consistent—the low was -103°F, the high 14°F. Pathfinder recorded this, but cared not. The mission was always rocks, rocks, and more rocks. Sojourner, after spending the better part of a week getting there, explored the area called Rock Garden, which was replete with interesting samples. Things went well until the rover got stuck on a rock, Half Dome, and again shut off automatically. It was too steep. But each time Sojourner did this in automatic mode, and had to be driven off of the obstacle, the teams back on Earth were learning. What this would mean for future missions was not yet quite clear, but gaining experience was the key, and it would bode well for future rovers.

Scientists continued to observe the smallest of details from the images returned: ongoing looks at the dirt under Sojourner's wheels—the soil-mechanics experiments—saw many layers of material, almost certainly deposited by water, and created an ever-expanding database of soil types. Farther out, the ground was covered by a layer of fine sand and drift, bright in color, indicating some differentiation of local soil conditions. This was in keeping with the idea of landing in a river-delta area.

In early October, communications problems returned. While signals returned from the errant lander computer indicated that the spacecraft was still functional, getting a meaningful conversation going was tough. Pathfinder was in trouble. The onboard battery seemed to be the culprit. It was losing capacity and was not only allowing the transmitter to get entirely too cold in the long Martian nights, but also failing to track time and date measurements. Low voltage and continuing resets of the computer were bedeviling JPL's plans.

If communications disappeared for more than five days, the rover was programmed to go into a contingency mode and, like a loyal dog, return to the base station (lander) and begin to circle it. This was designed to keep it from wandering too far afield or getting irretrievably hung up on a rock.

By mid-October it was becoming clear that the mission's days were numbered. On Earth, JPL engineers were testing identical hardware at a range of increasingly low temperatures in an effort to try to predict behaviors for the radio, but the results were not as useful as hoped. Still, the Pathfinder lander had outlived its planned primary mission of thirty days, and the rover had outlived its mission design of just seven days, but the ongoing sporadic failure of the radio was still a disappointment.

In three months of operations, Mars Pathfinder continued to refine the image of Mars as a planet awash in water during ancient times 3–4.5 billion years previous. However, the area surrounding Pathfinder appeared to have been dry and untouched by flooding for at least two billion years.

Sojourner's "nose," the Alpha Proton X-Ray Spectrometer, found some of the rocks confusing. There was far more silica in them than expected from studying Martian meteorites that had fallen to Earth. They appeared to be volcanic in origin, which argued for a highly active geological period in Mars's past. This rock type, called *andesite*, is typical of rocks formed by magma cooling in subterranean pockets, as opposed to the types of rocks found on some parts of Earth and on the moon. This latter type, called *basalts*, results from lava flowing onto the surface and cooling there in large sheets. But andesites are also indicative of active plate tectonics, which Mars did not appear to have, and are usually found at plate boundaries. Later observations from Mars Global Surveyor and other spacecraft indicated that Mars may indeed have experienced plate tectonics early on, with that activity ending far long ago (Earth's are still active).

By the end of its three months, Pathfinder had returned 2.3 gigabytes of data (by far the most accomplished in such a short period), over seventeen thousand images from the lander and the rover, performed sixteen detailed examinations of rocks, and sent back almost nine million bits of weather information. The team on Earth had gained valuable experience landing in an unorthodox fashion and driving a rover on Mars, which would prove invaluable for the next surface foray, the Mars Exploration Rovers. The understanding of the landing area had increased manifold, and modern electronics had been tested on the harsh and unforgiving surface of Mars.

Not a bad haul for a faster, better, and cheaper experiment called Pathfinder.

CHAPTER 15

ROBERT MANNING, MARS PATHFINDER

BOUNCING TO MARS

Rob Manning is a congenial and soft-spoken, if unintentional, folk hero. If you were a fan of the Mars Pathfinder website during that heady mission, you saw his bearded likeness all over the webcasts—calling out the numbers during the descent, announcing a successful touchdown, and throwing his head back with a fist pump when Pathfinder bounced to a stop. He was what might be termed the "principal cheerleader" as well as the chief engineer for the project, and has since taken these talents on to the Mars Exploration Rovers and the upcoming Mars Science Laboratory. When not busy with his projects at JPL, he pursues his varied hobbies, including jazz trumpet, in his Southern California home (just minutes from JPL) with his wife and daughter.

His introduction to the Pathfinder mission was a bit of institutional serendipity: "We have a paper here at JPL called the *Universe*, and it had an artist's rendering of this funny little mission, a rover, a very odd painting. I thought, 'JPL has no skill in this, it's been so many years since we've actually had to do something where we had to *land* on the planet.' . . . We hadn't really done a lander at JPL for many years. And while Viking was done at JPL, the lander was built at NASA Langley Research Center, who [*sic*] had since gotten out of the business of planetary exploration."[1]

Rob was puzzled that the lab would take on a Mars landing after all these years, yet at the same time, it was intriguing: "So I thought about this for some time. Soon I got a phone call from one of the people involved with this mission [spacecraft manager Brian Muirhead] who said 'I need an electronic whiz to work for me, someone who knows about computers and software.' I was an electronics- and software-systems engineer, so then we spent time in the JPL cafeteria and we hit it off, so he hired me as the chief engineer. I think he liked the fact that I wasn't super aggressive or controlling, and also that we really understood each other. That was around 1993. He taught me how to stuff a lot of electronics and complexity into a small little vehicle, and that was my charter: to work with the electronics team and the software team and the systems team. Over time it all grew together."

Mars Pathfinder and the Sojourner rover were built at JPL, giving folks like Rob a chance to have a very hands-on involvement with the project from planning to nuts-and-bolts to execution. This was a great way to have a personal stake in the mission, as well as save a lot of development money.

"You see, Galileo and Voyager, in fact nearly all of the flagship missions were built right here [at the lab], so JPL had a long history of hands-on work. I guess they just hadn't done one where they went through the atmosphere and actually landed, so that was going to be unique. The Sojourner rover was part of the Mars Pathfinder project, and that team was within JPL. So at the time I was actually responsible for Pathfinder. Now I didn't design Sojourner, but I was very involved with the design of Mars Pathfinder. Later I was also put in charge of the entry and landing [for the mission], as the main guy behind that. It was actually a trio of us, and we were the three legs that held the entry-and-landing part of it together.

"Now, for a long time Tony Spear, who was the project manager, was very concerned that we didn't have an entity to pull all those complicated systems together, and so he was looking to outsource it to another company. But it's a very complicated thing to outsource, and we didn't have much time before we launched. Launch was in late 1996 and [by this time it was] 1993, and I couldn't imagine writing a [specification] to tell people how to interface all those complicated things. How do you interface an airbag [landing system], for instance? So we finally talked Tony into letting us continue the process and building it here. Of course, we had lots of contractors working with us, so, for example, we designed the airbag architecture, but we didn't fabricate the airbag itself."

Outside contractors were brought in for key components—the airbags, the aeroshell, and the rocket motor. But the overall design and assembly of these components were to be in-house, and the people working on it worked hard—and enjoyed it.

"There was a lot of concern at NASA headquarters that connections for all those things that you need to build a vehicle like that might not exist here. JPL had the chops to build spacecraft, but nobody had all the chops to put the whole system together. So [in a leap of faith], Wesley Huntress, NASA's associate administrator for space science, called Tony Spear and said 'Tony, we would actually like you to implement this,' and so we did.

"So we started our own testing once the job came to us, [we] tried out different airbag concepts. We had to [figure out how] to [deflate] these airbags upon landing. On Mars the atmosphere is so thin that the air coming out of the holes vents supersonically, really fast. It doesn't do that on Earth, so the ability to test it on Earth, at least outdoors, was really lousy. Nonetheless we continued to try and get the venting to work.

"Now earlier on, JPL had proposed working with other vendors to do a nonairbag system, thinking that would be perceived as a more reliable and safer way to land. But the price tag for that was much higher. Viking had landed with throttleable engines, and the throttle is the coolest part. But coming up with a throttle that actually has the precision that you want, the dynamic range, was a very difficult and expensive proposition back in early seventies when it was developed for Viking."

One might think that this would be child's play by now, in the twenty-first century. But not so. As with the Apollo program, much of the brain trust and technical expertise of the 1960s and 1970s have been lost, as has the manufacturing capability. It was a problem for the Pathfinder team.

"The trouble is, there isn't a big call for these kinds of [rocket motors] in the space industry, and that throttling mechanism simply no longer existed by the time the 1990s rolled around. To make matters worse, all the people [who worked on it] were gone, and there was the sense that it would be very expensive to restore that technology. So [NASA] headquarters said, 'Keep it simple.' Remember that we're talking about a $150,000,000 spacecraft, all together, including the launch vehicle. That's actually really cheap, cheaper than a big movie at the time."

Amazing, but true. The motion picture *Titanic* cost $200 million to produce. Pathfinder was in line with the largest movie budgets of the time, and a bargain at that.

"So to keep it on budget, we thought the airbags would be the cheaper way to go. But it was also considered higher risk because we had never done it before. It had all these complexities that no one had ever dealt with; for example, how the hell to get out of it once we landed, how do you undress yourself on the surface of Mars without getting yourself wrapped up in knots and in tons of fabric? We had about two hundred pounds worth of fabric, and that's a lot."

The worries and the testing went on, seemingly ad nauseum. But eventually the kinks were worked out, sometimes by trial and error. It was all so delightfully ad hoc and low-tech, but it worked, and soon they were ready to go. One huge advantage of such a low-budget mission was that NASA's expectations were minimal.

"The great thing about Mars Pathfinder was that [it was predicated on] a single page of level 1 requirements. These were: head to Mars in the 1996 launch opportunity, land in 1997, deliver a rover, send back some pictures, good luck, and here was some money to do some science if you have some time. So the mission was to show that NASA could do things cheaply, efficiently, and effectively, and of course demonstrate that there are efficiencies in a faster, cheaper approach. Of course, this hadn't really been defined yet, because nobody ever agreed on what that meant."

This was the dictate of the then NASA administrator Daniel Goldin: faster, better, cheaper. It was not a bad idea in principle, but it had unfortunate consequences in many cases. However, Mars Pathfinder was one shining example of how it could work. One efficiency they achieved was to move the Pathfinder team, and the associated technical base, as close together as possible; it was almost reminiscent of the Mariner 4 days.

"We wanted to put all the team in one spot; we thought that was important. Now we couldn't quite achieve that, because of space availability issues at the lab at the time. But we got a majority of the core team members all in one spot. Communication was considered very important, teamwork was considered important, close management between the team was important, tight bandwidth between the team was very important; so we set it up to accommodate this. Also, don't overspecify things, it's okay to rely on oral communication as opposed to doing everything with legal documents, and above all keep it simple. That was our mantra.

"We didn't skimp on testing, though sometimes the test articles were low budget. But we didn't skimp on the tests themselves, so that was the overall model. The airbag was the perfect example of test-test-test. But there never seems to be enough people and resources to get the job done. . . . [T]here's never a sense of a forty-hour work week, it's almost a foreign concept. You can't predict scope with accuracy, yet you have to predict scope in advance to get the money, and there's a balance to what you can ask for and what you can get. You're expected to live within your means, so what happens is you get this variable called *homelife*, where you have to spend your time away from home all hours of night and on weekends."

But if their domestic routines were scrambled, most of the participants in the Pathfinder project seemed to replace that (temporarily at least) with an almost religious fervor in the mission . . . and they enjoyed it: "At Pathfinder I think we had a lot of fun because there was more team spirit, more sense that the team members in all levels of the project, even if they were a cog in the bigger wheel, could participate in a whole function of the thing by watching it. Part of it was the smallness [of the team] and part of it was the fact that there wasn't a lot of institutional pressure. It was a fantastic experience, everyone [who] worked at Pathfinder [whom] I know think of it as nothing but a fun experience."

Although the metaphor may be inaccurate, it is tempting to equate some of their fun with the toys at hand. Manning insisted, early on, that there be a way to test Pathfinder and the Sojourner rover in something remotely like a Mars-like environment close to the control room. This would help to avoid mistakes.

"Right above JPL's master control center was one giant room, and that room was for Mars Pathfinder. We had a sandbox in there, right above us. I had to do a lot of arm-twisting to convince people to bring in thousands of pounds of sandbox sand from down the highway in Monrovia. I wanted [the sandbox] right there so it could be the hub of our project."

We know the story of Pathfinder . . . the revolutionary bouncing landing, the deployment and travels of the plucky rover. But for Rob Manning, it was not just the spacecraft, but the people—and their interactions—that were fascinating.

"Mars Pathfinder had a very interesting attribute. Because it was a small project, representing a small number of people, each doing a large number of things, there's a lot of diversity in what people did for a living; if you were an electronics guy, you might find yourself doing airbags one day, or doing retracting tests, or software. There were a lot of unusual relationships, and I did all sorts of goofy things, mixing people up based on their skills, attributes, and interests. In a major business this is unusual because normally you would have people who have an agreement to deliver certain products, and when they're done with the delivery they're off the job. I didn't believe in that, because it's not the most effective use of people. It was always a pleasure working with the capability of people, and to me one of the continuing returns of Pathfinder was to see all these people do things that they didn't think they could do."

That said, it ultimately came down to the mission itself. It was a high-risk endeavor, trying lots of new techniques on the cheap. When it worked, the atmosphere at the lab was almost giddy: "I was the flight director during landing even, which really means that I knew when everything was supposed to happen, based on what the spacecraft was saying to us, which wasn't very much. We had a transmission delay of about eleven minutes at that time, so when we heard about it, it had either worked or it hadn't. During the landing, actually all the way down, I was giving a play-by-play on the net, with my headphones, and people were watching me as I was trying to interpret what I was hearing from [the tracking stations] . . . so it was really a lot of fun, and I was able to announce that it had landed, we had gotten a signal from Mars. It was really cool."

Of course, this had been simulated over and over. When dealing with long delays and highly complex automated systems, simulation is the engineer's best friend. "We practiced through our simulation setup, which was right next to our operation area. We would put an eleven-minute light-time delay between what was going on [in]our test area and [in] our control room. So even though we could walk between one room and the other in about ten seconds, it was like adding two hundred million kilometers of distance. We had practiced it so much that when we actually landed it felt fake! It didn't really click, just how real this was, until we got our first photo back, and the we realized that this was *Mars*. So we got to the surface of Mars and went through the whole process of getting the vehicle deployed, and then the rover's standing up, all within twenty-four hours. This was on the fourth of July. The first pictures came back in the evening, at about five o'clock Pacific time; that was our first view of Mars, including the picture of the rover. It was pretty darned exciting. There were also a lot of interesting stories about what had gone wrong, and surprises we saw. We had interesting anomalies after we landed; for example, our inability to talk to the rover. Because the antennae were crossed over, we found out that the signals weren't getting back and forth [between rover and lander] like walkie-talkies as well as they should have."

While the mission had flown under the radar for some time, when the landing was nigh, there was a ripple of delayed excitement—and concern—from NASA headquarters in Washington, DC: "We realized that Pathfinder was not going to be a little project. It might be little to us, but it was big to the outside world. We had so we lived [in the shadow of] of another mission, and much of NASA was just ignoring us. It wasn't until just before we landed that NASA upper management said, 'We better check this project out . . .' They knew about it of course, as they were funding us, but they were keeping their eyes away toward other challenges. They visited us just before we landed, and people like me told them why we thought this might work because they were not necessarily expecting success. They were *hoping* for success, but their confidence level was not very high."

It was no coincidence that, while the pressure from NASA and JPL might have been lighter than other missions, the pressure from the outside world was immense—for Mars Pathfinder was the first mission that was truly *live*.

"The good news is that we had a talented webmaster. Recall that at that time, 1997, the web was still fairly new. So he figured out how to use mirroring, so that the public could go to different websites, depending on where you lived, and you'd be redirected to a different server that had the Pathfinder website mirrored. This webmaster went around and actually talked to all these other companies about hosting us for free, and he was successful at it. As a consequence, when people actually went to the web to find out how the thing was going, not to just CNN, we had a huge spike in activity. We were astounded by how many were people were there. It was a different era in terms of bandwidth, we basically produced a huge spike in Internet traffic, which clogged things up for awhile. At that time it was completely unprecedented, but it gave people a taste of what might happen in the future. Our webmaster single-handedly figured out how to connect all these people, to make our website a success, and make it cool. He's one of the unsung heroes of Pathfinder."

And so is Rob Manning, along with hundreds of others at the Jet Propulsion Laboratory.

CHAPTER 16

MARS EXPRESS

On the Fast Track

After the failure of the Mars 96 mission, Russia must have felt "Mars Fatigue." Virtually every mission the Russians had sent off to the Red Planet had met with failure, from outright launch failure to trajectories askew (missing the planet entirely) to landers that failed upon touchdown. In any case, their record could be seen as one of failure perfected.[1]

However, this is too simplistic. Much has been learned along the way, and one only need look at their successes with the hell-hole that is Venus to see that the Soviet/Russian unmanned program has great merit. And, as the old Russian proverb says, "One beaten person is worth two unbeaten ones."

It is perhaps in this spirit that the Russian Federation approached its cooperation with the European Space Agency's Mars Express mission. Launched in 2003 atop a Russian rocket, the mission included many components of its own failed Mars 96 project. In a bit of technological cross-pollination, some of the technology on Mars 96 had come from Western Europe, so this was not as much an admission of need on the Russian part as a chance for continued cooperation. In addition to the European Space Agency's role, NASA joined the effort, bringing expertise in tracking and control to the table.[2]

Mars Express derived its name in part due to the extremely short distance the spacecraft had to cover at that particular launch opportunity: in 2003, Mars and Earth were closer than they had been in sixty thousand years. It would not do to wait for the next one. It should be noted that at under $200 million (US), it was also one of the cheapest Mars missions on record.

The probe consisted of two major components: the Euro-Russian Mars Express Orbiter and the British Beagle 2 lander. The lander was a small and fairly simple craft, designed to assess the usual components of the Martian environment—weather, landing-site geology and geochemistry—and even search for indicators of life. Unique to this craft were its origins: rather than the usual government-industry collaboration, Beagle 2 was born in academia. A professor at the United Kingdom's Open University, in association with the University of Leicester, promoted the idea, eventually drawing in two other universities and four industry partners. The final result was a worthy craft, a small "clamshell" probe with a manipulator arm, designed to land via parachute and airbags, not unlike the Mars Pathfinder before it.

The successful orbiter followed traditional concepts, with a central body flanked by solar panels. The instruments onboard were designed to meet an increasingly familiar set of goals:

A spectrometer working in both visible and infrared wavelengths called OMEGA would determine surface mineral composition.

Another spectrometer in the ultraviolet and infrared wavelengths, called SPICAM, was specifically designed for sensing the composition of the atmosphere.

A radar altimeter called MARSIS would seek subsurface water.

A Fourier Spectrometer to measure atmospheric temperature and pressure.

A high-resolution stereo camera could photograph surface features.

And various radio and energy-sensing experiments were also onboard.

Mars Express was also equipped, as were Mars Odyssey and the Mars Reconnaissance Orbiter, to be a relay for NASA's other Mars landers and rovers.

As the craft neared Mars, the Beagle 2 separated to continue along its own path, bound for the planet's surface. It eventually made it, but not alive. So far as can be gleaned from the data, some part of the landing system failed and Beagle crashed.

The Mars Express orbiter was luckier, attaining orbit around the planet in late December 2003. This mission eschewed aerobraking; a small rocket engine was used for slowing and to allow orbital capture. As a result, the craft went into a highly elliptical orbit, 185 miles from the surface at its lowest point and 6,280 miles at its highest. Not ideal for orbital work, but far simpler (and safer) than pursuing a circular orbit.

Notable accomplishments of Mars Express are many. The poles were studied, resulting in a measurement of 15 percent water ice and 85 percent carbon dioxide there. Methane and ammonia were sensed in the atmosphere; this is noteworthy because neither would last very long in the Martian air, so a source of continual replenishment must exist. And that source could be active volcanoes, hydrothermal vents or . . . *living things.*[3]

Of course, water was again spotted, both as current ice deposits and as areas indicative of a wet past. Intensive atmospheric investigations were made, helping to identify the rate at which the air is thinning on Mars. Hydrated (water-altered) minerals were observed at the poles, and similar rocks were spotted in Valles Marineris, which continues to narrow down the time scale of the aqueous episodes of Mars. The idea of a wetter Mars in the distant past, followed by a drier, harsher planet in more recent epochs, as seen in the geological record, was strengthened. Auroral displays were observed above areas of strong magnetic activity. Finally, the lumpy gravitational field was observed and recorded.

The MARSIS instrument allowed for a more direct look beneath the surface of the planet, revealing yet more indications of subsurface water. MARSIS was further able to probe the intricacies of the polar caps, giving a better idea of the total mass of water ice there. The southern ice cap alone has a maximum depth of over two miles, and if melted, it could cover the entire globe to a depth of about thirty-five feet! Finally, a fascinating frozen mass of water was found in the Elysium region near the Martian equator—a place it really had no right to be. And it is young by geological standards—only about five million years old.

Not bad for a seemingly dry, dead world.

The mission of Mars Express has been extended numerous times and continues to this day. The probe returns a continuing stream of images and data from Mars and serves as a valued complement to NASA's own orbiters as well as an outpost of European scientific endeavors. Results from this mission have added greatly to the ever-growing knowledge of the Martian environment and its processes. In particular, the puzzling observations of stray methane in the Martian atmosphere have many researchers intrigued.

Europe and its partners will return to Mars with the ExoMars probe soon, possibly as early as 2016. With good planning and a dash of luck, some of these questions may be resolved.[4]

CHAPTER 17

A LAUGH IN THE DARKNESS

THE GREAT GALACTIC GHOUL

It came from nowhere, hid in the darkness, attacked things earthly and then retreated once again.

It was the Great Galactic Ghoul, the monster that hides somewhere between the orbits of Earth and Mars in our solar system, whose sole purpose is to devour unwary spacecraft and plunge earthbound scientists into despair. And the ghoul is good at its job.

Almost forty spacecraft have headed off to Mars. Nineteen have arrived intact and functional. NASA's score: thirteen out of twenty. Better than the unfortunate whole (which is heavily weighted toward Soviet-era failures), but grim nonetheless. Had the Apollo program suffered such losses, few astronauts would have stepped up for future missions.

But Mars is not the moon, and traveling to the Red Planet (even ignoring the added complication of taking human beings) is far more difficult and time-consuming. Mars is a killer of probes, a consumer of human capital. And this is perhaps fitting for the God of War.

The origins of the ghoul's name are uncertain; some credit it to a *Time* magazine reporter from the 1960s, others to various personnel within JPL and NASA. However, one recent mention occurred in 1997 referring to the partial blockage of the Mars Pathfinder's rover ramp by the deflated airbags (the problem was later resolved, and the ghoul was cheated out of lunch).[1]

Over the years, the ghoul has gobbled up many machines, primarily but not exclusively from the early days of space exploration. A partial list includes:

1960

Mars 1960 A, USSR—Launch failure
Mars 1960 B, USSR—Launch failure

1962

Sputnik 22 (Mars 1962 A), USSR—Broke up shortly after launch
Mars 1, USSR—Lost contact before Mars flyby
Sputnik 24 (Mars 1962 B), USSR—Failed to leave Earth orbit

1964

Mariner 3, US (Mariner 4 succeeded)—Launch shroud failed to deploy properly
Zond 2, USSR—Lost communication ninety days before reaching Mars

1969

Mars 1969 A, USSR—Launch failure
Mars 1969 B, USSR—Launch failure

1971

Cosmos 419 (Mars 1971 A), USSR—Launch failure
Mariner 8, US (Mariner 9 succeeded)—Launch failure
Mars 2, USSR—Orbiter succeeded, lander crashed on Mars
Mars 3, USSR—Successful landing on Mars, then operated fifteen seconds on the surface before failure

1973

Mars 4, USSR—Missed Martian orbit, flew by into deep space
Mars 5, USSR—Lasted nine days in Martian orbit before failure
Mars 6, USSR—Some data returned before crashing on Mars
Mars 7, USSR—Lander separated early and missed Mars entirely

1988

Phobos 1, USSR—Lost contact in space
Phobos 2, USSR—Lost contact in space

1992

Mars Observer, US—Lost contact in space

1996

Mars 96, Russia—Launch failure

1998

Nozomi (Planet-B), Japan—Failed to orbit Mars
Mars Climate Orbiter, US—Erroneous aerobraking maneuver, broke up in Martian atmosphere

1999

Mars Polar Lander, US—Crashed on Martian surface
Deep Space 2 (affiliated with MPL), US—Crashed on Martian surface

. . . and so forth.[2] Surveying the list, one might wonder (a) why the Russians bother to continue trying at all and (b) what the heck they did wrong. Analysis of the once-competing US and USSR programs yields mixed conclusions, but it is apparent that Mars exploration has been largely an American game. And as noted, even JPL has its troubles.

December 11, 1998: many folks at work around the United States were just beginning to ponder Christmas vacation activities. But at JPL, most eyes were on the launch of the Mars Climate Orbiter, or MCO. A successful launch would result in Martian orbit late in the following year, and reams of new data about the weather and surface conditions on Mars. MCO was also to act as a relay for the next landers scheduled to set down on Mars. Anticipation was high.

Then, on January 3, 1999, the counterpart to MCO, the Mars Polar Lander, departed Cape Canaveral also bound for Mars. It would descend to the surface of Mars shortly after MCO and begin tandem operations. Life was good.

The ghoul awoke, smelling opportunity, and in late September 1999 reached out and wrapped his taloned hand around the Mars Climate Orbiter. On the twenty-third, communication was lost between JPL and the spacecraft. This was at a critical time when the craft was to begin aerobraking maneuvers to slow it down and circularize its orbit, as the Mars Global Surveyor had successfully done before it. The probe entered the Martian atmosphere at an improper angle and, so far as is known, broke up upon encountering the denser-than-expected air.

The ghoul retreated into the darkness, its simple task complete. Hearts were broken on Earth, and a light rain of thin metal parts burned up high in the Martian sky.

JPL mourned the Mars Climate Orbiter . . . and looked forward to recovery with a successful Mars Polar Lander.

The ghoul smiled a knowing smile.

JPL and NASA scrambled, meanwhile, to determine the cause of the failure. The press releases were somewhat terse (but straightforward) at first, as there was not a lot of information available to the authors: "Early this morning at about 2 a.m. Pacific Daylight Time the orbiter fired its main engine to go into orbit around the planet. All the information coming from the spacecraft leading up to that point looked normal. The engine burn began five minutes before the spacecraft passed behind the planet as seen from Earth. Flight controllers did not detect a signal when the spacecraft was expected to come out from behind the planet."[3]

"We had planned to approach the planet at an altitude of about 150 kilometers [93 miles]" said Richard Cook, project manager for the Mars Surveyor Operations Project at JPL. "We thought we were doing that, but upon review of the last six to eight hours of data leading up to arrival, we saw indications that the actual approach altitude had been much lower: it appears that the actual altitude was about 60 kilometers [37 miles]. We are still trying to figure out why that happened."[4]

Huh? The probe went into a low and fatal orbit, in this day and age? After all the Apollo flights, after the many successful Mars orbiters? After the complex Viking missions, for god's sake? After aiming spacecraft at planets throughout the solar system, swinging around four or five of them to reach Jupiter, Saturn, Neptune, and Uranus? Is this the same JPL that tracked Pioneer 10 all the way out of the solar system and into deep space beyond?

Lamentably, yes. And it was a simple human error that precipitated the deadly mistake.

A few days later, the painfully inward-looking announcement came out: "The root cause of the loss of the spacecraft was the failure of translation of English units into metric units in a segment of ground-based, navigation-related mission software. . . . [T]he failure review board has identified other significant factors that allowed the error to be born, and then let it linger and propagate to the point where it resulted in a major error in our understanding of the spacecraft's path as it approached Mars."[5]

The announcement went on to elaborate on a multilevel failure. Too few personnel, a new management design, the handoff of the spacecraft from the design, build, and launch team to a new, multi-mission operations team. In short, JPL was trying to do too much with too little. It had always been known for this ability, but had crossed a threshold. And a large part of this was, quietly, placed at the feet of NASA top leadership and the quest for "faster, better, cheaper."

But there was one more Mars-bound mission currently en route. It was the Mars Polar Lander, and its success would redeem the lab's reputation. Perhaps the ghoul's appetite had been sated, for now.

On January 3, 1999, the lab tried again. Mars Polar Lander left the cape aboard a Delta II rocket to begin the long coast to Mars. All went well as the craft left the influence of Earth and headed off on a complex trajectory toward Mars. The lander, which measured ten feet by four feet, would descend to the Martian polar area to search for water. Two parasitic craft were affixed, Deep Space A and Deep Space B. These were impactors, and they would descend ahead of the lander to hard-impact the surface and do science work of their own, penetrating up to a yard into the icy soil. The mission would be one for the record books, and could revolutionize our view of the Martian polar regions.

The probe carried the usual array of cameras, a laser-sounding instrument that could detect aerosols in the atmosphere, a robotic arm with digging scoop, a gas analyzer (with eight tiny ovens) to determine amounts of oxygen, water, carbon dioxide, and other life-bearing elements, and a microphone to send home the sounds of Mars. It was a wonderful package for exploring an exiting environment. And attached to it was abundant excitement, and that indefinable attribute of human endeavor, hope.

On December 3 of the same year, the craft entered the Martian atmosphere. All was going well. The lander plummeted into the thin veil of Mars, soon to deploy a large parachute to slow the fall. The usual communication blackout occurred as it screamed through the thickening air, headed for the planned soft landing . . .

And the controllers waited, and waited. And they waited some more. All too soon, it was apparent that something had gone wrong.

Increasingly anxious attempts were made to establish communication with the lander. The Mars Global Surveyor orbiter was pressed into service in an attempt to photograph the planned landing zone and find the craft, but to no avail.

Subsequent analysis by the Failure Review Board came up with a likely hypothesis. The vibrations emitted by the deployment of the landing legs may have caused the onboard computer to assume that the craft had landed, and then it did what it was supposed to do—shut down the descent engines. Unfortunately, the ship would still have been well over one hundred feet above the frozen ground, and if the hypothesis is correct, would have slammed into the cold soil at deadly speed.[6]

Whatever the actual events, the mission was concluded in a rude and depressing fashion. Repeated communication attempts failed to rouse the lander, and the mission was assumed to be lost. Coming on the heels of the loss of the ill-fated Mars Climate Orbiter, it was another black eye for unmanned space exploration, and the blame fell to JPL. Tears were shed, heads hung low, and the team members who had been so fastidiously assembled to operate this exciting mission were prematurely disbanded and sent off to other projects or back to their home institutions. The Mars Polar Lander joined the annals of lost spacecraft, and the ghoul licked its lips once more.

It would be irrational to feel cursed, but more than one "JPL'er" could not completely abandon the idea. But a more relevant assessment of the root cause of this failure, coming just under three months after the previous debacle, prompted Thomas Young, the chairman of the Mars Program Independent Assessment Team, to proclaim that the program "was underfunded by at least 30 percent." Insufficient staffing, insufficient testing, and insufficient review had taken a fatal toll once again.[7]

And, once again, "faster, better, cheaper" had proved to be anything but.

You don't hear the ghoul mentioned at JPL much anymore, though it can still incur a nervous chuckle when mentioned. More to the point, when discussing some of the failures from the past, are memories of poor decisions from the top and questionable implementation within the ranks. Like any large bureaucracy, NASA has angels and a few demons scattered throughout. But on the whole, the agency (and while managed by Caltech, JPL is a part of NASA) does amazing work with ever-tightening budgets. If the failure rate even began to approach that of the early days—the 1960s—public (and more to the point, the dreaded congressional) outcry would probably slam the lid closed on the entire operation.

As it is, funding is desperately hard to come by. But the men and women of JPL soldier on, driven not by stratospheric salaries or visions of corporate power and grandeur, but by the desire—no, the *need*—to investigate the great darkness beyond, to discover the mechanisms of the universe, to set foot—whether flesh or robotic—onto new worlds and find the microcosmos within. And so, despite the setbacks, JPL moves forward—with a revamped management structure and a revised rule book—to explore. And the ghoul will have poor hunting in the new millennium.

But he will be there . . . waiting.

CHAPTER 18

2001

A MARS ODYSSEY

Mars and Earth both have elliptical orbits, one inside the other, so at varying times (about every two years), Mars and Earth are in "opposition," when they make their closest approach to one another. Due to this, there are favorable launch opportunities every two years for Mars-bound spacecraft. This is a driving force behind scheduling for the Mars program at JPL, because a delay of even a few weeks in development, planning, building, or testing can cause a mission to be delayed for two more years. It is a headache common to all the "Martians" at the lab, as they often refer to themselves.

With the turn of the millennium, JPL and NASA were still recovering from two very embarrassing mission failures. Internal and external reviews had illuminated many failings within the development and the management structure of the program, and these revelations were hard to swallow. Heads rolled, teams were restructured, management methods were reevaluated, testing procedures were strengthened, hardware was reexamined, and perhaps most important, budgets were scrutinized. "Faster, better, cheaper" was deemed, arguably, to have been a fallacy and was quietly retired, though money was still a tremendous challenge.

And throughout, the Mars exploration program moved forward. Incredibly, even with all the shuffling and restructuring, a few missions stayed on track. And the next one to launch, the relatively inexpensive Mars Odyssey, was ready to go in 2001.

The spacecraft was built by Lockheed Martin. The aerospace contractor had proved to be a capable and willing partner, unusually cooperative in unmanned space efforts, an arena where government/contractor relations can get sticky. Previous partnerships with NASA and JPL had gone well.

The primary goal of Mars Odyssey would be to search for water from orbit. To do this, the spacecraft would carry two principal instruments: one, called THEMIS (Thermal Emission Imaging System), would image the planet in infrared, allowing scientists to map Mars in temperatures instead of traditional visual wavelengths. The infrared images could then be aligned with images taken in visible light, and the correlation of visible landforms with areas that radiated stored heat at night would provide a better understanding of the mineralogical makeup of much of the planet. Toward this end, Mars Odyssey also was capable of taking traditional images as well.

The other was called HEND (High Energy Neutron Detector), which would identify elements in the Martian environment, specifically in the first few inches of soil. Here, planetary scientists would be looking for hydrogen, an indicator of water and water ice. Between the two instruments, it was hoped to clarify where the moisture might be on Mars and in what concentrations, among other things.

A third instrument was called MARIE (Mars Radiation Environment Experiment), which would measure radiation in the Martian orbital path. This was not only of interest in strict science terms, but would also assist in the planning of eventual crewed missions to the planet.

The overall spacecraft was about the size of an upright refrigerator, with a boom extended out one end (which held the gamma-ray spectrometer sensor) and solar panels out to two sides.

Mars Odyssey left Earth on April 7, 2001, aboard a Delta II rocket and successfully headed off toward the Red Planet. As the rocket sped toward Earth orbit, the distance from Earth to Mars was about seventy-eight million miles, but due to the course the spacecraft would follow to reach the its goal, Mars Odyssey would travel over 285 million miles.

Once within Mars's wispy embrace, aerobraking was again used to circularize the lopsided orbit inherent in missions utilizing a smaller rocket (the technique saved almost 450 pounds of fuel, which is a huge amount in Mars-bound launches). After the braking rocket fired, dropping Mars Odyssey into that lopsided orbit, aerobraking continued for almost three months until the ellipse became a circle that was proper for surface mapping to begin. This required not only circularizing the orbit but also adjusting it to an almost north–south orientation, also known as a *polar orbit*, which has been used for most post-Viking orbital missions. With this, the planet rotated perpendicular to the orbit of the spacecraft, allowing it to repeatedly photograph almost every part of the surface as the planet slowly turned below.

And then, in late February 1992, Mars Odyssey got to work. The major risk milestones had been passed, and things seemed to be going well. The folks at JPL breathed a bit easier. Mars Odyssey, it seemed, would not disappoint.

You see, lessons had been learned since the multiple failures of the 1990s. Most of the flight systems (i.e., most things likely to fail or malfunction) were now redundant and had backup units. It was almost as if JPL had taken a page from the manned spaceflight playbook, in which as many systems were doubly and triply redundant as possible.[1] This was, in some ways, now applied to Mars Odyssey.

The brain of the spacecraft was a radiation-hardened version of Apple's Macintosh® processor of the time, the Motorola Power PC® chip. With 128 megabytes of RAM and three megabytes of other storage, it was hardly a powerhouse but would do the job.

As data moved from here out to the rest of the spacecraft (and back), an ingenious parallel design was used. Computer cards enabled specific tasks, and were placed in double rows downstream from the processor, allowing for 100 percent backup capability. The only parts of the computer not backed-up were the main processor and a (then staggering) one gigabyte storage card for imaging.

Finally, the flight software, which was carried onboard as opposed to the commands later sent up from Earth, had more sophisticated "fail-safe" routines written in and had frequent self-checks. If something went wrong, the craft would immediately enter a "safe" mode and begin troubleshooting the situation. This had always been a part of deep-space software, but had been beefed up following the reviews of recent losses.

By late March 2002, Mars Odyssey was sending back images that were then posted online almost immediately by the JPL team. It was the second Mars mission to offer the public such immediate feedback, and it had been pioneered by the impressive online presence of the Mars Pathfinder mission. The images were spectacular even in their raw state, and doubly so once enhanced.

And in an ongoing quest, Mars Odyssey was looking for life on Mars.

Of course, as an orbiter, it did not have the luxury of a dirt scoop and an analytical lab like Viking did, nor the ground-based mobile capabilities of Pathfinder. But with the growing understanding of the nature of Mars, the Odyssey team had been able to focus its instruments on the investigation of water present on Mars. Where water was found, past or present, there could be life, past or present.

The strategy was to look at the surface, and shallow subsurface, environment of Mars for both hydrogen, indicative of water, and various mineral types, indicative of past water. The probe would also be able to spot hot springs now suspected to possibly be on Mars. With patience, a holistic picture of the planet would eventually come to light. The machine simply had to function long enough to allow these results to emerge . . . and in this, Mars Odyssey would shine like none before or since.

By April 2002, Mars Odyssey had already allowed planners to select landing sites (from many dozens of candidates) for the upcoming Mars Exploration Rovers, which would leave Earth in just a few months. The images being studied were critical to making an informed choice, and the results had been immediate and gratifying. There was a high level of confidence in the selected sites.

Concurrently, visible light images had been combined with earlier shots from Mars Global Surveyor to examine some gullies that appeared suspiciously like drainage channels. It was soon realized that a few had been caused by *recent* melting of water snow. This was a breakthrough, as it provided proof that water was still "running" on Mars, something long suspected but until now unproved. The key idea was that ice was melting underneath snow packs, and the mass of ice above prevented the water from flash-evaporating in the thin atmosphere as it flowed out to create the gullies.

On Earth, a few little erosion channels might not cause even a moment of excitement, as we are used to seeing such things constantly and easily in our own dynamic ecosystem. But these gullies had first been observed on Mars in 2000 from Mars Global Surveyor images, after countless hours of painstaking investigation of thousands of pictures. They seemed to occur only on the colder, pole-facing "shadow" side of some craters, and as such, indicated that a special set of circumstances was contributing to their formation. This colder, shaded location allowed snow to accumulate and remain across an entire Martian season, so that melting could occur gradually, allowing the water to seep out and create the features observed. Here was the explanation, the "smoking gun," that so many had been waiting for. For a Mars scientist, it was nirvana.

Soon another result came in, supplying a broadly opposite picture of another part of Mars. When the region called Ganges Chasm was investigated, a huge deposit of the mineral olivine was discovered. Olivine is soluble in water over time, so this large an area of the mineral indicated a long dry period in the region. The infrared imaging was also allowing planetary scientists to develop a far better idea of mineralogical distribution, lava flows, and soil types. All this was derived from observing thermal or temperature differences between one area and another, as seen in daytime and then at night when the more slowly cooling areas (indicative of differing soil and rock types) were clearly visible in this invisible spectrum. Not only do different kinds of rocks cool at different rates, but sand and sedimentation (dirt, pebbles, boulders) of a given type of rock cool faster than a solid mass of the same material. So the temperature measurements showed broad swaths of geological and erosional formations, allowing the paint-by-numbers map of Mars to be filled in, for the first time, with some authority.

These results did much to redeem JPL in the public eye, as well lift the spirits of those who labored there and at affiliated institutions. But there was more to come. The gamma-ray spectrometer was showing the planetwide distribution of ice under the surface, demonstrating far more water than even optimistic projections had predicted. This helped to answer the question of where all the water had come from to form the many huge, water-created features seen across the surface, and ruled out some of the more far-fetched hypotheses. Mars was now confirmed as having a *currently* living (if less so than Earth's) environment, and the Red Planet was far from the dead place it had appeared to be so long ago in the fuzzy Mariner 4 pictures. There was clearly water—if not gold—in those hills out yonder.

Look at it this way: not all the visual images coming back were of much higher quality, and in some cases less so, than the previous Mars Global Surveyor. But the thermal information was almost three hundred times better. So until now, scientists had been forced to deduce geological and environmental conditions from images that could see nothing smaller than a schoolbus—not as useful as they desired. But the new and highly detailed thermal information from Mars Odyssey showed what was below the sand, what existed in large areas invisible to the naked eye, painting a larger and better-defined map of the surface composition.

But wait, as the Ginsu® salesman said, there is more.

The instruments also showed large areas of bare rock where the excessive dust and sand found all over Mars had been scoured away, as well as large deposits of rubble at the base of hillsides and mountains. This was a dead giveaway that weather was hard at work on Mars, reinforcing the notion that, while dry and dusty, there was a vibrant meteorological system at work on Mars.

As the days grew colder in the Martian winter of the northern hemisphere, the expected dry ice layer appeared. But when it retreated, a dense layer of permafrost—water ice in the soil—was quickly apparent below. Again, water, water, everywhere, even if it was frozen. This was good news for those seeking possible habitats for some form of life on Mars.

This was remote science at its best. In many cases, incomplete answers from Mars Odyssey were fleshed out by data from Mars Global Surveyor and vice versa. The two spacecraft were working in tandem—each according to its strengths—to develop a clearer, planetwide picture of Mars. And this gave scientists something else they had been lacking heretofore: context.

The picture of Mars that was emerging was an intriguing one. Water ice was found widely distributed across the planet. The concentrations lean out nearer the equator; in the polar regions, one might find a half pound of water per pound of soil, or 50 percent water by weight. Closer to the equator, this falls to 2–10 percent. There were exceptions though. Arabia Terra, an almost 2,000-mile-wide equatorial desert, and other equatorial exceptions showed indications of large masses of water. Mars was much more complicated than most had thought.

It is worth noting that the water masses are, in some areas, projected to be about half a mile deep. But the instrumentation on these orbiters was able to measure only about a yard down, so anything beyond that is theoretical. However, this deep water-mass projection does come close to resolving the mystery of the missing water on Mars as regards total mass.

One more shock awaited. The MARIE experiment, designed to measure radiation in the Martian orbital environment, presented a surprising picture: the amount of radiation, the kind that might put future astronauts at risk, was at least double that of Earth. As a control, measurements were taken aboard the International Space Station during roughly the same period. The experiment had previously been used on both the station and the shuttle, but this was the first time it had left Earth orbit. MARIE measured both solar radiation and galactic cosmic rays. The result? Astronauts spending extended periods in the Martian orbital environment would need extra protection from radiation. The experiment ended when a spike in solar flares appeared to have overloaded the device.

But Mars Odyssey had one more trick up its sleeve. By the time it had reached its first Martian year of operation—687 Earth days—it had been pressed, as planned, into being a relay station for the newly landed Mars Exploration Rovers. A full 75 percent of the early images from the rover Spirit were relayed through Odyssey, and the orbiter continued to render assistance for the life of Spirit (until it ceased function), as it continues to for the rover Opportunity.

And Mars Odyssey has earned one more distinction. In late 2010, it broke Mars Global Surveyor's record as the longest operational spacecraft at Mars: 3,340 days. It is still going strong and continues to provide data, imaging, and relay capability for other missions.

While it may not look much like the Energizer Bunny®, Mars Odyssey just keeps going, and going, and going . . .

CHAPTER 19

DR. JEFFREY PLAUT

FOLLOW THE WATER

Jeffrey Plaut started his college education studying music composition but was unable to maintain his allegiance to Bach and Beethoven. He can often be found on the porch of his small home at the foothills a few miles east of JPL in Sierra Madre, a small hamlet that resembles Mayberry as much as a Southern Californian suburb. From his back patio, with a view of the nearby San Gabriel Mountains, he is as likely to be listening to Jimi Hendrix as Joseph Hayden. He is not able to spend as much time as he would like with his wife and two young daughters due to his long and rewarding involvement with Mars Odyssey. From his home he recalls those formative college years:

"There were several different things I was interested in, and one of them was music, and I did composition, I played the piano, a whole variety of stuff . . . and I pursued that, but I always had an interest in math and science, and I sort of kept that going in the background while I was doing my music major. I was taking some math courses and astronomy courses senior year, and I took a class called Planetary Geology. It was a graduate course, so I was somewhat out of my league, having not done any geology up to that point. I managed to make it through the course, and I got kind of excited, I wrote a term paper that my professor enjoyed about the moons of Jupiter and the possibility of life there. So I graduated as a music major, but the geology sort of stuck with me, and after a couple years I decided to look for a career in it and got into a program at Washington University in St. Louis, and that has one of top programs in planetary geology, and the adviser was tied in with JPL, and that's how I eventually got into JPL."[1]

Once on the track to JPL, there was no turning back. He had been bitten by the Mars bug.

"I came on as the deputy project scientist, and the guy who I succeeded, Steve Saunders, was something of a mentor for me here at JPL, and I guess he liked having me as his right-hand man, so he brought me onto this project; then he retired and I got moved up to the project scientist position. For the last twelve or thirteen years, I've been focusing mainly on Mars, and I've also worked on two other Mars orbiter projects, [the Mars Reconnaissance Orbiter] and Mars Express."

But Mars Odyssey was not initially planned as simply an orbiter. It was something far more complex, in the vein of the Viking missions of 1976: "This project actually consisted of both an orbiter and a lander that we were going to fly to Mars. The orbiter was going to land first, and the lander, along with a small rover similar to the Pathfinder rover, was going to land. So the orbiter would have two jobs, one would be to relay the data, as we do now for the Mars Exploration Rovers, and the orbiter would also make measurements around the planet as well as handle the lander's data.

"That mission unfortunately never flew. We were well in development for the lander part of the Mars 2001 mission, and when we had the twin failures on the Mars 1998 project, which were of course the Mars Polar Lander and the Mars Climate Orbiter.

"So, half a step in, they said that the Mars 2001 Lander was extremely similar to the Mars Polar Lander, which failed, so they said, 'Let's still hold on to that, we still have the orbiter.' [T]hey understood the problems, and it really wasn't any fundamental problem with the orbiter itself, so, we went on and did the 2001 mission and the lander was put on the shelf . . . and eventually got resurrected as the Phoenix Lander. [It had the] exact same hardware, [was the] exact same unit that was supposed to go to 2001, which eventually went to a 2007 mission and was very successful."

In fact, the Phoenix Lander became the first successful landing above equatorial Mars, and the mission, though brief in comparison to Mars Odyssey, was wildly successful. But that's another story.

One instrument would set Mars Odyssey apart from all previous orbiters, and its name was THEMIS: "It's a unique instrument [that] makes images using infrared vision. Its detector is taken straight out of night-goggle technology, and it sees a part of the spectrum where there are diagnostic spectral signatures of certain minerals that appear on Mars that are not easily detected with other [spectra] on instruments. And another thing unique about it is that it will create global maps of Mars, almost 100 percent coverage of Mars, so basically it makes an image of the planet's temperature. You can see how the surface gets cooled down during the night. [The] rockier areas stay warmer during the night while dusty areas cool down faster, so the camera can really tell us a lot about the terrain on Mars and its texture. The resolution is really incredible, about 100 meters per pixel."

Odyssey was well equipped to make history. "I think we're already at the point where we can look back and see what's historical and what really this mission has achieved. There are two different areas. First is the discovery and mapping of ice in the soil. The onboard instruments made unequivocal observations and maps of hydrogen in the subsurface of Mars down within the first couple of feet, and we saw both polar regions and down to about sixty degrees north and south latitudes. [This is] what you might call the Arctic of Mars . . . just shot through with ice in the soil. There really was no way for anybody to make that measurement before, and make the maps, until Odyssey came along. This provided the target for the Phoenix lander, which set down within this arctic circle. Besides the scraping with its robotic arm and those investigations, the soil was blown away by the descent boosters of the lander, and it uncovered ice right underneath it. So historically speaking, that might be the biggest mark that Odyssey has made and will be remembered for.

"The whole theme of this Mars exploration program is to follow the water and to understand the possibility of life on Mars. Clearly all life that we know of needs water in its cells and its environment to survive. So it's always been the major goal of the mission to understand the role of water in both the history of Mars and also the evolution of Mars today. It's as if to say, 'where can we find water [and] the ice,' and to be able to localize a map, and ultimately have it confirmed for the [landing site]. That was a huge step, to follow the water and touch it with the Phoenix lander. We have several other plans to send landing craft to Mars. None of [the others] are going to these icy terrains, but I think, ultimately, we will go back to some of these icy areas, maybe to find a place where there might be a hospitable environment for some kind of little microbe."

But the Mars Odyssey mission was not all guts and glory: "I think the most difficult period during the mission was about two years after we arrived, which was around October 2003. There was a series of huge solar flares, and that resulted in a kind of radiation or magnetic storm at Mars, and it just clobbered our spacecraft. It actually killed off the MARIE instrument; the sensor just measured this radiation and choked on it. The storm also set Odyssey into a safe mode, which is a good thing if a spacecraft's in trouble. It goes into a safe state, where it's not actually required to do a whole lot, but we did lose contact for a time. When we got it back, we saw that it had been rattled, and we had to improvise, press the reboot button and do a complete hard reset of the computer. That is a bit stressful. But other than that we have been very fortunate."

So what lies ahead for Mars Odyssey, currently the longest serving spacecraft at Mars? "We are going to go for as long as we can! We are already way beyond our prime mission. One thing that helps is that we served this relay function, we're continuing to do that for the [Mars] Opportunity Rover, [and] hopefully in a year or so, when the Mars Science Laboratory arrives at Mars. We still have good science with the instruments we have around, [and] as long as we have fuel, we still should be able to continue to operate. We just might have another ten years, if we don't run out of fuel or funding."

These two factors, fuel and funding, are the great nemeses of robotic exploration of the cosmos. Fuel is a fixed quantity once the craft leaves Earth, but continued funding is something that dedicated explorers like Jeffrey Plaut worry about every day.

CHAPTER 20

TWINS OF MARS

SPIRIT AND OPPORTUNITY

In the first years of the new millennium, spurred by the success of Mars Odyssey, JPL seemed to regain its institutional confidence. Things were working again, and Mars seemed within reach in a way not seen since Viking.

The next step after the spectacular success of Mars Odyssey, which continued to operate and send back information vital to future mission planning, was a set of dual rovers. These would be an evolution of the Mars Pathfinder mission: similar in design but an exponential leap in scope and ambition. They were the Mars Exploration Rovers (MER).

These twin rovers, which built on knowledge gained from the successes (and limitations) of the Mars Pathfinder rover, Sojourner, were built at JPL. In general terms, the orbiters tended to be built by outside contractors (Lockheed Martin preeminent among them) while the rovers were built at the lab by internal staff. The design and fabrication of Pathfinder had been exemplary; the Mars Exploration Rovers would outshine even that.

Each completed spacecraft would weigh in at about 2,400 pounds, with the rovers themselves tipping the scales at about 408 pounds. Rather than depending on the landing stage as a relay for the radio transmissions back to Earth (as Sojourner did), MER would utilize spacecraft already in orbit around Mars, the Mars Global Surveyor and Mars Odyssey probes. It was an ingenious and carefully planned perfection of the capabilities of JPL assets on and in orbit around Mars. The rovers were also capable of communicating directly with Earth, but the orbiters offered a superior conduit for communication.

The rovers were both far larger and more robust than Sojourner, but with a similar overall design. These too used solar panels for power, and each would arrive on Mars sitting inside a lander shielded by metal petals. The lander itself would follow a flight profile similar to Pathfinder's, and would employ an almost identical landing scheme, right down to the beach-ball cocoon and the multiple-bounce arrival. Why mess with success?

But while Sojourner had provided a few short weeks of successful operations within sight of its lander, MER would range far and wide over long and active missions. To provide a maximum return on investment, the instrumentation had been beefed up as well.

For starters, the rovers were loaded with cameras. There was a panoramic camera, mounted on a mast about five feet high, to image the surrounding terrain. On the same mast was a navigation camera, with a wider field of view. This one operated in black and white and for driving and navigation purposes. Below this was a mirror for the Thermal Emission Spectrometer, which helps to identify promising rocks and soils for closer investigation. Finally, there were four more black-and-white cameras, two up front and two at the rear, for hazard avoidance. Their sole purpose was to assist in keeping the rovers out of trouble.

One more imager made up the visual complement: the Microscopic Imager, which would take extreme high-resolution close-ups of the rocks and soils being investigated by the arm.

The instruments for scientific investigation took their cue from Sojourner and expanded on this theme. These were mounted on the same robotic arm as the Microscopic Imager, which gave the rover even more reach. There was an Alpha Proton X-Ray Spectrometer (an improved version of the APXS on Pathfinder) that could identify the elements of the rocks that the rover would stop and "sniff." Another device was the Mössbauer spectrometer, used to investigate iron-bearing rocks and soils.

Less high-tech but still useful was the oddly named "RAT," or Rock Abrasion Tool, which would dust off or, if necessary, grind down the surface of rocks to be examined. This allowed for a clean, fresh surface to test with the various devices. And last but not least, there was a collection of magnets, to pick up any ferrous material from the RAT or from the environment at large, which the Mössbauer spectrometer would more closely analyze. This device is particularly adept at identifying iron-bearing minerals that other devices may not be able to "see" when present in small amounts. It can sense the magnetic properties of samples and potentially identify materials formed in hot and wet environments. If there was a downside of this particularly valuable device, it was that a thorough reading took up to twelve hours. But the rovers would have plenty of time.

Altogether, it was a neat and tidy little science package, which owed a lot to the successes of its predecessor, Pathfinder's Sojourner rover.

The landing zone for Spirit, the first rover to descend, had been carefully selected. It would have to be smooth enough that the airbag landing method would work. It had to be low enough in elevation that there would be sufficient atmosphere to pass through for slowing of the heavier rovers to occur. It had to be near the equator, and not so potentially dusty that the solar panels would become disabled. Over 150 candidates were considered; for this first landing, the final choice was a large crater named Gusev.

About fifteen degrees south of the Martian equator, Gusev was named after a nineteenth-century Russian astronomer. It is almost one hundred miles wide and geologically speaking is a transition zone between the ancient highlands to the south and the smoother, younger plains to the north. And entering the crater from the southeast is a 550-mile-long meandering valley called Ma'adim Vallis, which appeared to have been massively eroded by water at some time in the distant past. The hope was that, since the water appeared to have emptied into the crater, the floor might have layers of sedimentation that could be explored.

It was a cleverly selected site, decided upon by a combination of acquired knowledge, deduction, and detective work. Its promise seemed clear.

Spirit arrived in a fireball on January 3, 2004, entering the Martian atmosphere at over 12,000 mph. Once again, airbags were used to cushion the blow of the high-speed entry, and the machine bounced a couple miles before settling into its final landing spot.

Upon arrival, the craft deflated the airbags, unfolded its petals, and took a preliminary image, again, just as its predecessor Pathfinder had. On the ground in Pasadena, those glued to the monitors were ecstatic. It was just what they had hoped for.

To the untutored eye, the flat expanse with a few rocks would seem desolate. But for a rover control team and the associated scientists, it was heaven. The rover had a perfect-looking surface to traverse, a wide selection of rocks to explore, and an open horizon to seek. And, perhaps most important, it was different than that encountered by either the Viking landers or Pathfinder.

For the first week, the rover sat in place and surveyed its surroundings. Nearby was a depression about thirty feet wide, soon dubbed Sleepy Hollow (features near landing sites never remained anonymous for long). It was either a wind-worn hole or a meteor crater; either way, it offered immediate access to subsurface geology. It was the proverbial hole in one.

But before this would be explored, a number of more symbolic gestures were to be made. First, there was a plaque aboard the rover, which was dedicated to the astronauts lost in the space shuttle *Columbia* accident with a moment of silent observation. Then a DVD, stored atop the spacecraft, was imaged by its camera. It was a funny moment: here was a DVD, with a little Lego®-style robot printed on it, and all of it held in place by what appeared to be Lego building bricks. It was all part of a sophisticated but seemingly simple effort to reach out to kids, to budding young scientists. This is something that JPL has done exceedingly well, especially since the Pathfinder mission with its huge Internet component. This DVD held the names of *four million* people, part of the "Send Your Name to Mars" outreach project, along with other student messages to future human or alien explorers. But not to forget the science, simple magnets had been attached to the edges of the disk to allow students around the world to study how much ferrous metal was contained in windblown dust. It was clever beyond measure, and fun to boot. Before too many years have passed, the first of these students will be entering careers in space science, many inspired in part by this simple gesture. It was an unusual moment of marketing genius by NASA, and it cost next to nothing to accomplish.

But now, it was time to explore. To the transmitted strains of Bob Marley's "Get Up, Stand Up," Spirit flexed its robotic muscles and prepared to roll off its platform, a week after arrival. But before it went a-strolling, more images were sent home, these with the infrared spectrometer. After the rousing success of infrared imaging by Mars Odyssey, MER had taken a similar instrument along for the ride. Once again, materials invisible to the naked eye could be seen in the surrounding terrain, but this time, it was at ground level. Besides helping to identify the composition of the rock, the instrument also spotted dusty areas to be steered around. It was the best thing since putting wheels on Mars!

First the front wheels, then the rear ones, were extended. This was a carefully observed process. The suspension of the machine had acquired its DNA from the Sojourner rover, using the same "rocker-bogie" arrangement of swinging arms and six wheels.[1] Once upright, a cable near the center of the rover had to be cut to allow it to move free of the lander. NASA had long ago learned the danger of plugs and connectors coming loose on spacecraft, so, since the dawn of the space age, when spacecraft connected by a wire to another craft or the ground needed to separate, this was accomplished with explosives and knifelike guillotines. While this may sound extreme, it works well in practice, and the failure rate has been very low. *Bang* went the pyrotechnic charges, a blade swung and the wires were cut, never to rejoin. Spirit said good-bye to its lander. It was the last of 126 pyrotechnic charges fired since the launch of Spirit almost a year earlier. Nobody said exploring space is simple . . . or quiet.

Slowly, so very slowly, Spirit embarked on the first step of its long drive to places unknown. Driving on Mars was a slowly evolving science. It had been done only once before, with the Pathfinder mission. And that rover, Sojourner, had gone only a few dozen feet away from the lander that was its link to home. Spirit and Opportunity would range far and wide, communicating (it was hoped) directly with JPL's orbiting spacecraft as they passed overhead every ninety minutes or so. But still, despite the wealth of experience, despite the somewhat-autonomous hazard-avoidance software onboard, and despite a long delay of outgoing radio commands (and a similar delay on verifications from Mars), the rovers needed to be driven by humans on Earth. It was the old mile-long drinking-straw analogy, the item through which mission controllers had to labor to make things work on Mars. It was not as simple as jumping into a car and driving off.

There were many layers in the process and many skills to be learned. First, the surrounding terrain was extensively imaged by Spirit to give the drivers a sense of place and a taste of the road ahead. Then, through communication with the Mars Odyssey orbiter overhead, Spirit got a better fix on its actual location. Finally, via the infrared images, it was possible to map out the difficult terrain and any "sand traps" in the surrounding area.

In the few years prior to the landing, JPL's "rover drivers" had been in a sort of extraterrestrial driver's ed class. Called Field Integrated Design and Operations, or FIDO, it involved taking a fac-

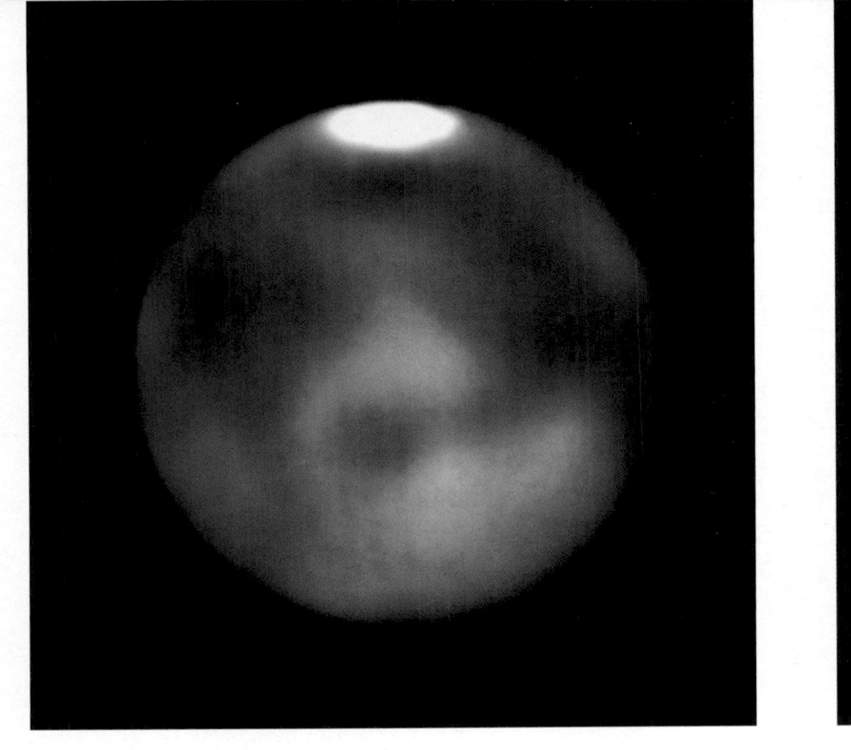

MARS THEN: This composite of telescopic images of Mars is about as good as it got through telescopes prior to the transmissions from JPL's early Mars probes. From Earth-based telescopes, the image swims due to the motion of the air in the atmosphere, and it is easy to see how early astronomers could have seen features that did not exist. *Courtesy of iStockphoto.*

MARS NOW: The Red Planet as seen from the Hubble Space Telescope at a distance of fifty-four million miles. Features as small as twelve miles across are discernable. The north polar cap is visible at the top, and the dark region below this is called Acidelia. To the center left of this is the Valles Marineris area, home of the largest canyons in the solar system. Morning clouds are visible to the left limb of the planet. *Courtesy of NASA.*

HORUS: A depiction of the god most frequently associated with Mars in Egyptian mythology. *Courtesy of Ad Meskens.*

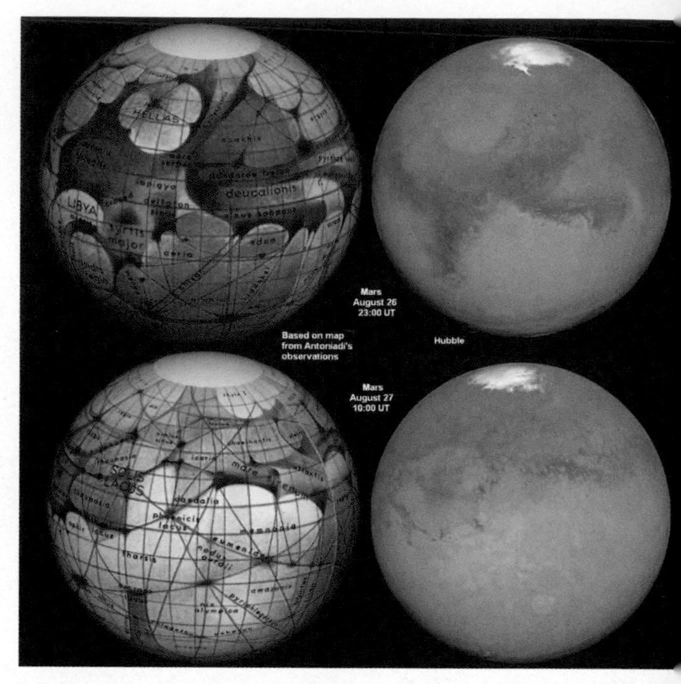

FACT VS. FANTASY: On the right are Hubble Space Telescope images of Mars. On the left are modern renderings of the "classical" Mars of Giovanni Schiaparelli, Percival Lowell, and Eugene Antoniadi, the last of whom also observed and sketched the planet but later doubted the reality of the canals. By looking at these modern comparisons, we can see that not all the canals and structures were purely illusory—some aligned with major surface features. *Courtesy of Selden E. Ball Jr. and NASA.*

MARINER 4: In 1965, NASA/JPL sent the first unmanned probe to Mars, Mariner 4. The four arms radiating from the central body are the solar panels for providing power, and the flaps at the ends (used only on this mission and later to Mercury on Mariner 10) were intended for steering via solar wind deflection. The louvers on the sides of the main structure can be opened and closed to facilitate cooling of the internal electronics. *Courtesy of NASA/JPL.*

BY-THE-NUMBERS: This early image from Mariner 4 is not yet processed by the slow computers of 1965, so a scientist is coloring by hand, following the digital data (on cut strips of paper) to get a sense of the landscape on Mars. *Courtesy of NASA/JPL.*

MARINER 9: The family resemblance waning, Mariner 9 is clearly a departure from previous Mariner spacecraft. Larger, heavier, and far more capable, this spacecraft entered orbit around Mars instead of merely swinging by. It returned images for about one year, revolutionizing our understanding of Mars. This mission made it clear that wind and possibly water forces had shaped—or were still actively shaping—the surface of the planet. *Courtesy of NASA/JPL.*

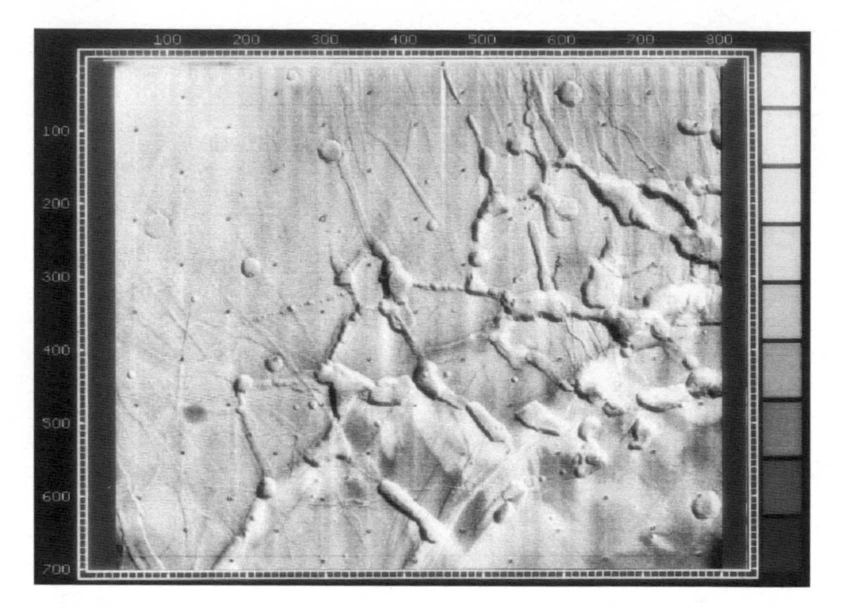

A REVOLUTION: This image from Mariner 9, obtained after weeks of dust storms cleared, shows the Noctis Labyrinthus region, at the western end of Valles Marineris. Some powerful natural force was clearly at work here, shaping the surface of the planet. The "dead Mars" group had something new to think about. *Courtesy of NASA/JPL.*

VIKING 1: By 1975, Viking 1 was on its way to Mars. This spacecraft was a quantum leap over previous unmanned efforts to any planet. The lander is enclosed in the "aeroshell" at the top, while the orbiter is seen below. The rocket engine, the nozzle of which is seen at the base of the craft, had to be large enough to decelerate the huge mass into a stable Martian orbit without aerobraking. It was a brute-force approach to landing on Mars. Bigger, heavier, and far more complex than any previous mission, Viking was a leap of faith into the unknown, which allowed the United States to become the first country to successfully land on Mars. *Courtesy of NASA/JPL.*

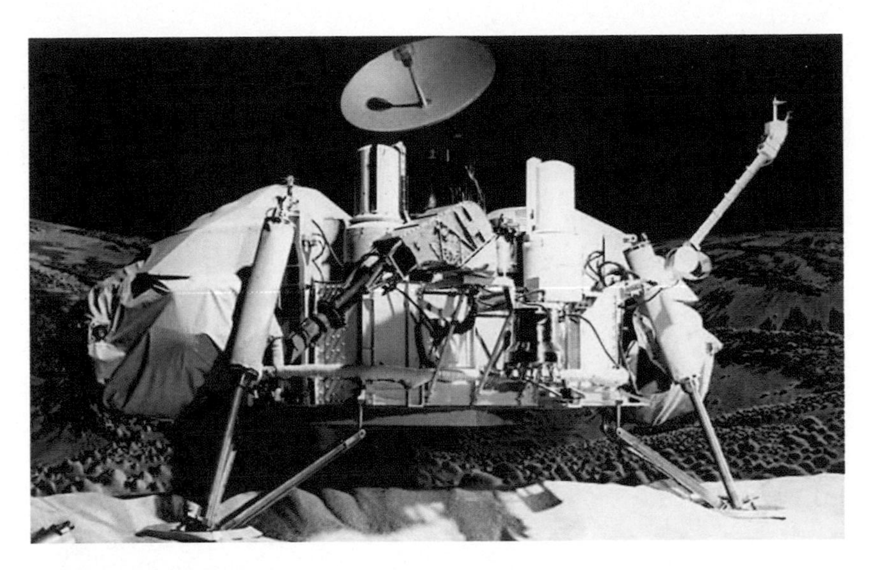

VIKING LANDER: This mock-up of the Viking lander clearly shows the basic structure of the first successful Mars lander. The spherical structures to left and right are the fuel tanks. The tan device to the upper center is the retractable sampler arm. The two "can-shaped" structures on the front top deck are the high-resolution cameras. Finally, the arm to the right is the meteorology boom. *Courtesy of NASA/JPL.*

MARS FROM VIKING 1: An image taken as Viking 1 approaches Mars. Across the center, in unprecedented detail, can be seen Valles Marineris, the globe-straddling chasm that scars the planet. *Courtesy of NASA/JPL.*

FIRST ON MARS: This is the first color image returned from Viking 1 of the horizon of Chryse Planitia. The Viking imaging team had worked hard, and as quickly as possible, to get the color right, but later efforts changed the overall tone substantially. Nonetheless, this was a big moment, as, after years of Soviet attempts to reach the surface of the Red Planet first, Viking succeeded where they failed. *Courtesy of NASA/JPL.*

SUNSET ON MARS: This Viking 1 image depicts a cold and bleak sunset on Mars. The banding seen in the sky demonstrates a limitation of remote imaging circa 1976. *Courtesy of NASA/JPL.*

SAMPLES, ANYONE? The Viking lander's meteorology boom crosses the frame to left. In the soil of Chryse Planitia can be seen small trenches dug by Viking's sampler arm. While much good data about the soil and overall environment of Mars was gained from the Viking mission, the results regarding microbiological life, while generally regarded as being a negative finding, are still being debated to this day. *Courtesy of NASA/JPL.*

THE PLAINS OF UTOPIA: Viking 2 phones in from Utopia Planitia. At the top of the frame is the dish for communication with Earth. To the left, the American flag is seen on the propellant tank cover, and in shadow on the right is a color chart. This chart, identical to the one on the Viking 1 lander, allowed image technicians back at JPL to accurately calibrate the color of the images returned by the cameras onboard. *Courtesy of NASA/JPL.*

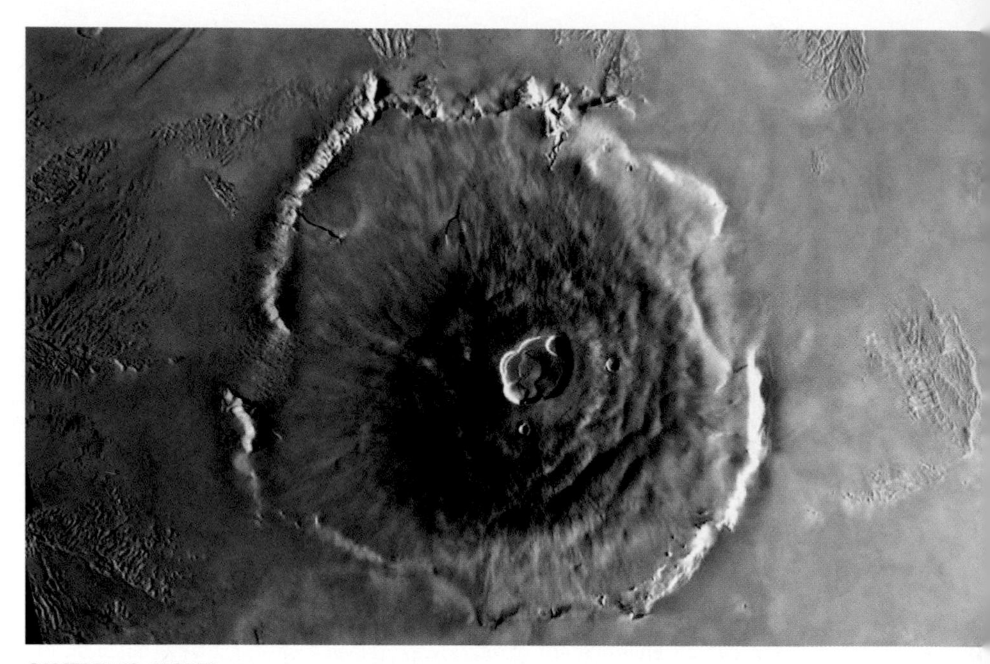

OLYMPUS MONS: The largest volcano in the solar system, Olympus Mons is seen here through the eyes of the Viking 1 orbiter. Prior to its discovery by the Mariners, this region was known on telescopic maps as Nix Olympica. But once the dust cleared during the Mariner 9 flight, it became clear that the feature was an enormous shield volcano. *Courtesy of NASA/JPL.*

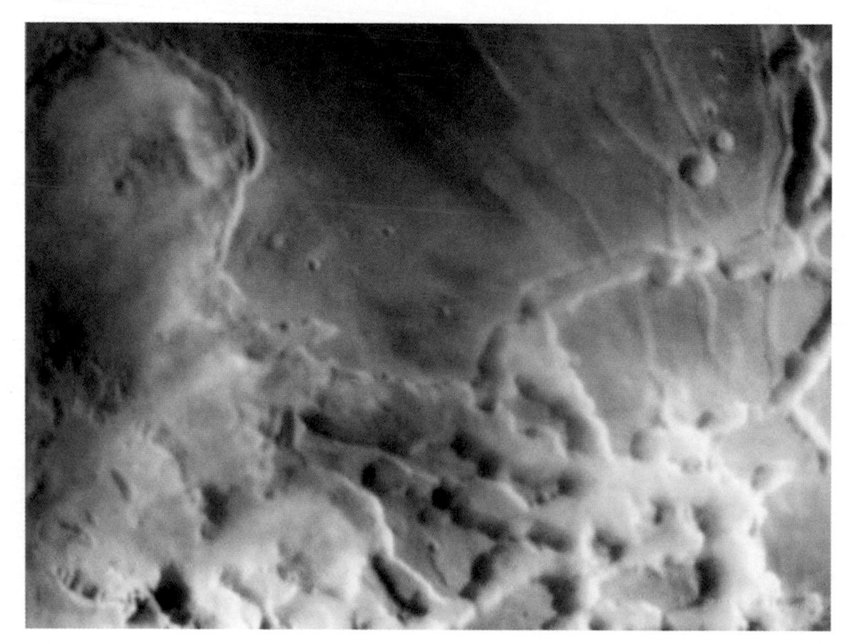

NOCTIS LABYRINTHUS: Another image sent home by the Viking orbiters, this area clearly shows the effects of erosion on the Martian surface. This image is a close-up view of the same territory shown in the Mariner 9 image of Noctis Labyrinthus; the improved image quality is obvious. The misty white seen in the valleys is exactly that—early-morning fog. *Courtesy of NASA/JPL.*

THE VALLEY OF THE MARINERS: The Viking 1 orbiter also investigated the planet-girdling feature Valles Marineris. This, the largest valley in the solar system, represented an unprecedented look at the mechanisms responsible for the huge depression. The specific area shown here is called Ophir Chasma. *Courtesy of NASA/JPL.*

VIKING MERCATOR MAP: This global Mercator projection of Mars was made from a mosaic of Viking orbiter images. It was the first complete map made from orbiter images. *Courtesy of NASA/JPL.*

WINTER AT UTOPIA: This 1979 image from the Viking 2 lander shows a thin coating of water ice on the rocks nearby. It is an incredibly thin layer, thinner than a tissue, yet it lasts over three months out of the long Martian year. *Courtesy of NASA/JPL.*

ALPHA AND OMEGA: This stunning, computer-augmented view comes from data provided by the Mars Global Surveyor, the mission that heralded America's return to Mars twenty years after the Viking missions. And in a completely unintended bit of serendipity, one of the craters on the terminator is Gale Crater (top center of image, with a mound at its middle), ninety-six miles across and the intended landing site of the Mars Science Laboratory in 2012. *Courtesy of NASA/JPL.*

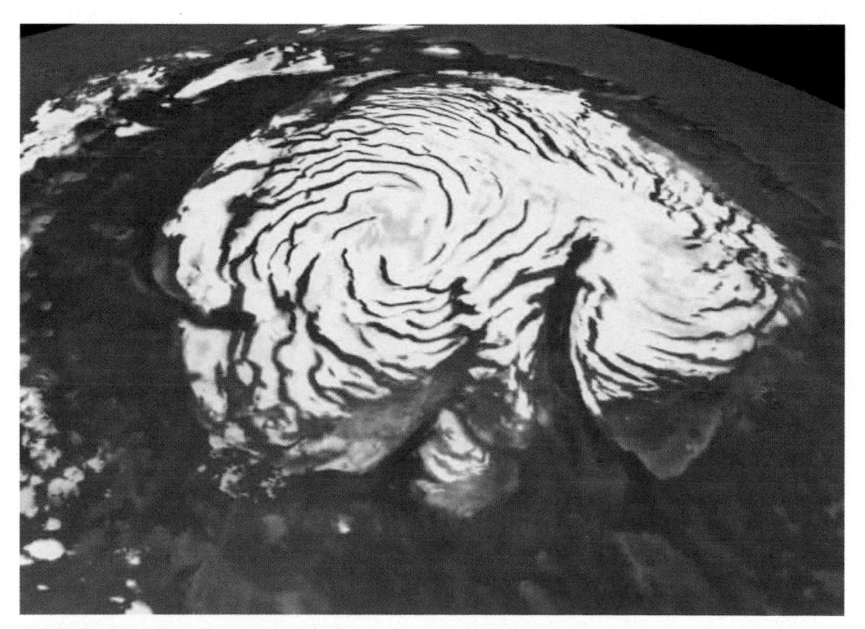

NORTH POLE: This Mars Global Surveyor (MGS) image depicts the north pole of Mars. MGS and other mapping Mars orbiters are placed into polar orbits, that is, orbits that run north to south instead of equatorially, to provide the maximum coverage of the planet as it rotates below them. The central portion of the ice cap is about 620 miles across, expanding in the winter and receding in the summer. *Courtesy of NASA/JPL.*

PHONE HOME: These radio-tracking antennae are at Goldstone in the high desert of Californi and are a part of the Deep Space Network (DSN). Each dish is 110 feet across. JPL runs the DSI in conjunction with NASA and its international counterparts in Madrid, Spain, and Canberr (Australia) to provide continuous twenty-four-hour coverage of the various missions it supports *Courtesy of NASA/JPL.*

SNIFF, SNIFF: The Mars Pathfinder rover, dubbed Sojourner, used its front-mounted probe t sniff out chemical compounds. The rock it is investigating is called Yogi. These simple chemica analyses could take many hours or even days to complete. Sojourner was the first rover on Mars and it succeeded beyond its designers´ dreams. *Courtesy of NASA/JPL.*

N THE YARD: The Sojourner rover investigates a rock not far from its base station, the ᴾathfinder lander, seen at the bottom of the frame. During its short life, Sojourner never ventured ar from its home base, as the lander was its link to Earth. The later MER rovers would sever that ie. *Courtesy of NASA/JPL.*

FAMILY PORTRAIT: The MER rover is seen here behind its predecessor, the much smaller Pathfinder rover. The MER rovers would go farther and longer than anyone imagined possible with the second arrival, Opportunity, continuing through today. Their success owed much to the experiences gained from Pathfinder. The 2012 Mars Science Laboratory is similarly larger than MER. *Courtesy of NASA/JPL.*

AT THE CAPE: Not Canaveral, but Cape Verde inside Victoria Crater. This bluff became a primary point of interest during the MER mission, as it exposed a vast swath of Martian geological history. This image is from October 2007, 1,329 days into the mission. *Courtesy of NASA/JPL.*

BLUEBERRIES: A close-up view of the infamous "blueberries." Found in various locations throughout Meridiani, the landing site of Opportunity, these hematite formations are strongly indicative of the past presence of water. On average, the spherules are about the size of a BB pellet, or about an eighth of an inch across. *Courtesy of NASA/JPL.*

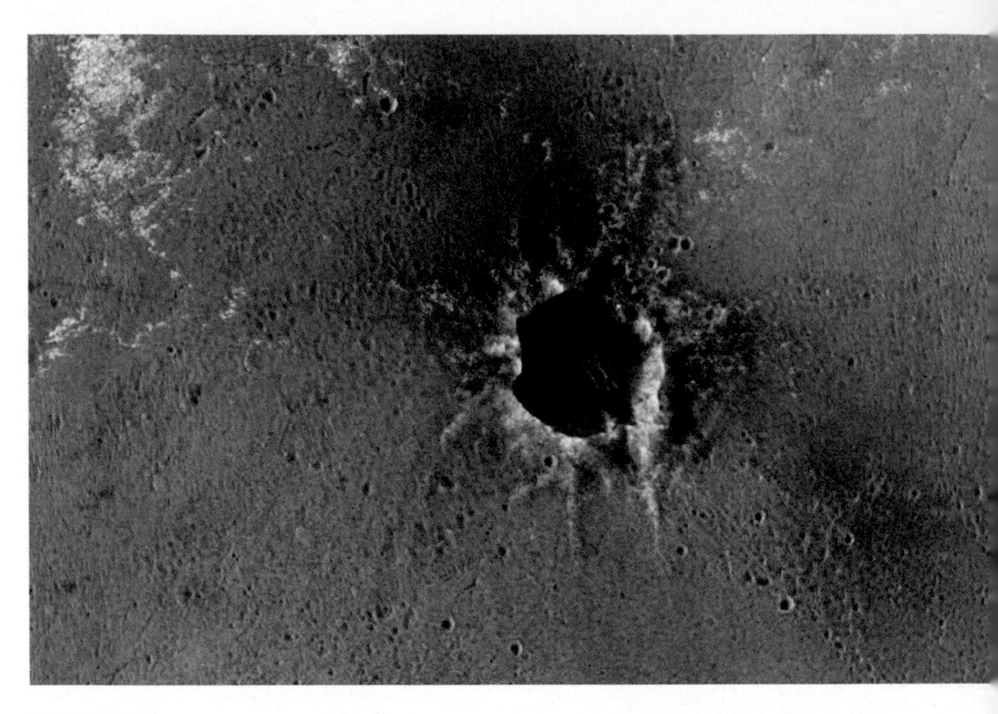

ENDEAVOR: This is an orbital image of Endeavor Crater, one of the target sites of Opportunity. The tracks of the rover as it approached the crater can be seen to the left. The fiery appearance of the crater rim is due to the exaggerated color of the image, often used by JPL scientists to enhance the detail visible within the image. *Courtesy of NASA/JPL.*

THE PLAINS OF MERIDIANI: A stunning panoramic shot taken by Opportunity during its travels across Meridiani Planum. The region is not so different from the Mojave Desert in California, where engineers spent time evaluating exploration methods. *Courtesy of NASA/JPL.*

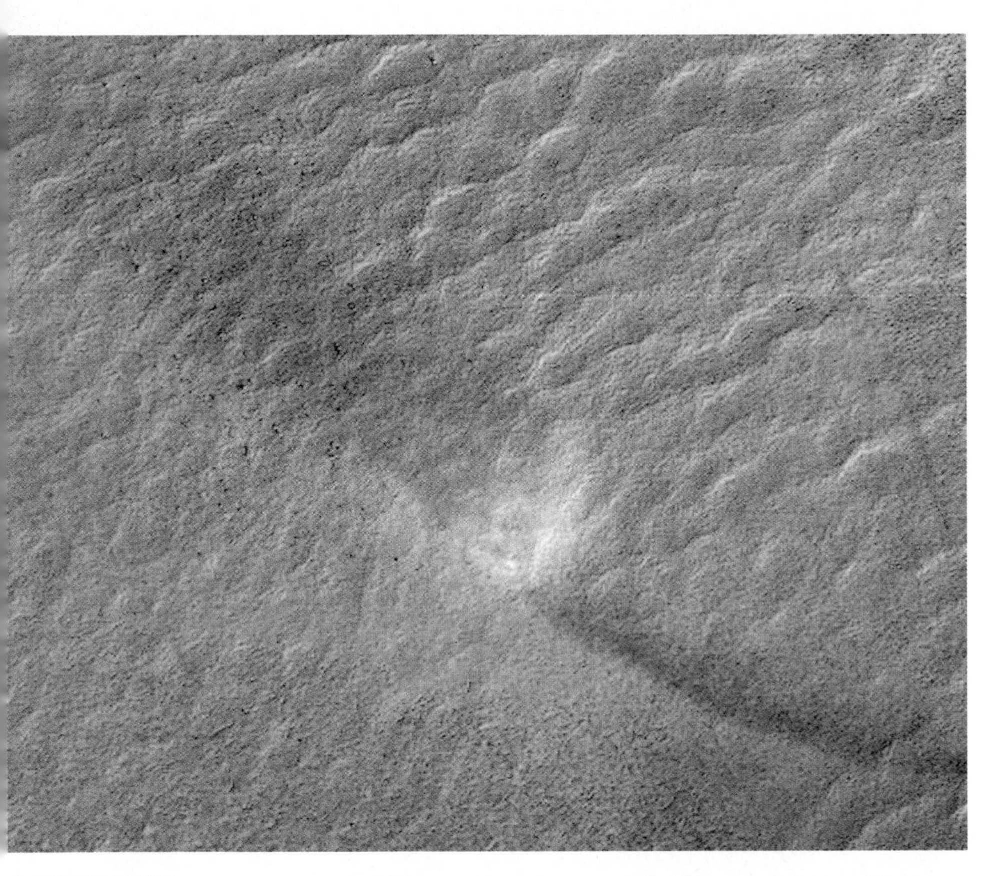

A DUSTY DEVIL: A dust devil as seen from orbit by the Mars Reconnaissance Orbiter. These are the same kinds of wind-driven phenomena photographed by the MER rovers, though this one is considerably larger than most. Its shadow can be seen to the left, and the track it leaves can be seen to the lower right. *Courtesy of NASA/JPL.*

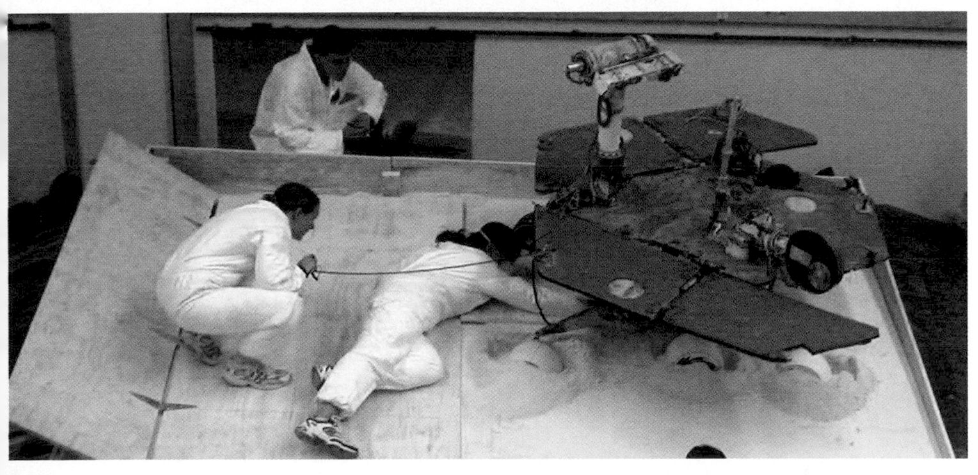

STUCK: Technicians try to re-create the conditions that entrapped the rover Spirit to devise a solution to its woes. Eventually, malfunctioning wheel motors, sand-trap conditions, and a failing power supply doomed the first of two MER rovers on Mars. *Courtesy of NASA/JPL.*

VICTORIA CRATER: This magnificent image was taken prior to the traverse of this crater by the MER rover Opportunity. Almost 2,500 feet across, Victoria was far larger (and far more treacherous) than the previously visited Endurance Crater. Opportunity spent almost a year exploring the interior of the crater. *Courtesy of NASA/JPL.*

TRACKS: Tracks left by Opportunity while driving in "autonomous" mode. Although much experimentation had already been done, mission planners were still very cautious before turning over control of the MER rovers to their onboard computers. In the end, however, the software (with a few updates) was very successful. *Courtesy of NASA/JPL.*

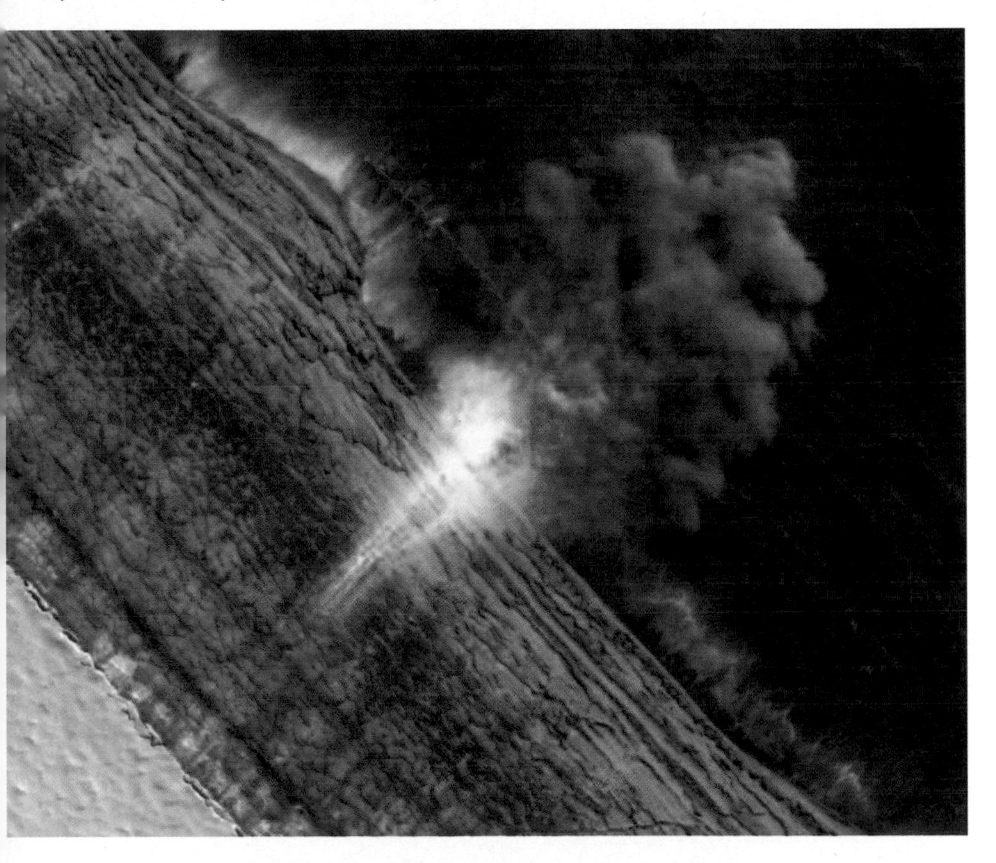

AVALANCHE: For what was once considered a "dead" planet, Mars can get very busy. This image, snapped by the Mars Reconnaissance Orbiter, shows a real-time view of a massive avalanche about six hundred feet across. The cascade may have been caused by melting ice or high winds. *Courtesy of NASA/JPL.*

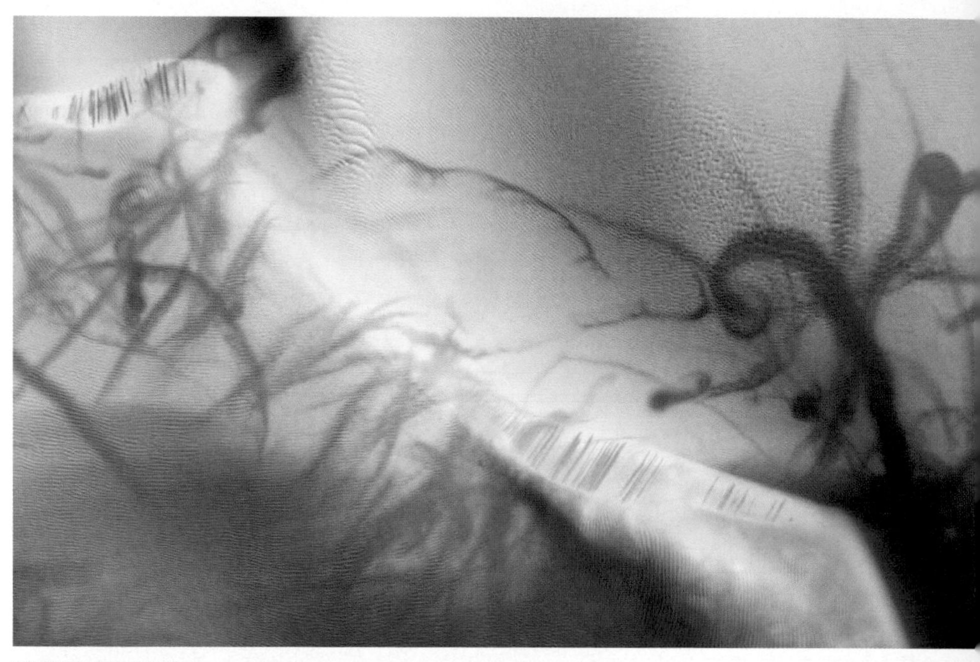

MARS TATTOO: Created by winds, these swirls of darker material overlie the lighter dunes beneath. The Mars Reconnaissance Orbiter captured this image in August 2009, almost three years after its arrival at Mars. *Courtesy of NASA/JPL.*

HAUNTING BEAUTY: Another image from the Mars Reconnaissance Orbiter shows a region known as Arkhangelsky Crater. The dunes, blue, overlie a harder surface, brown. The colors are exaggerated to enhance detail. This type of sand dune is called a *barchan*, with a gentle slope enclosing a steeper face with two horns. *Courtesy of NASA/JPL.*

MARS RECONNAISSANCE ORBITER (MRO): This artist's rendering shows the MRO spacecraft at work around Mars. The dish, seen at the top, is the largest flown on a Mars mission, at about ten feet across. *Courtesy of NASA/JPL.*

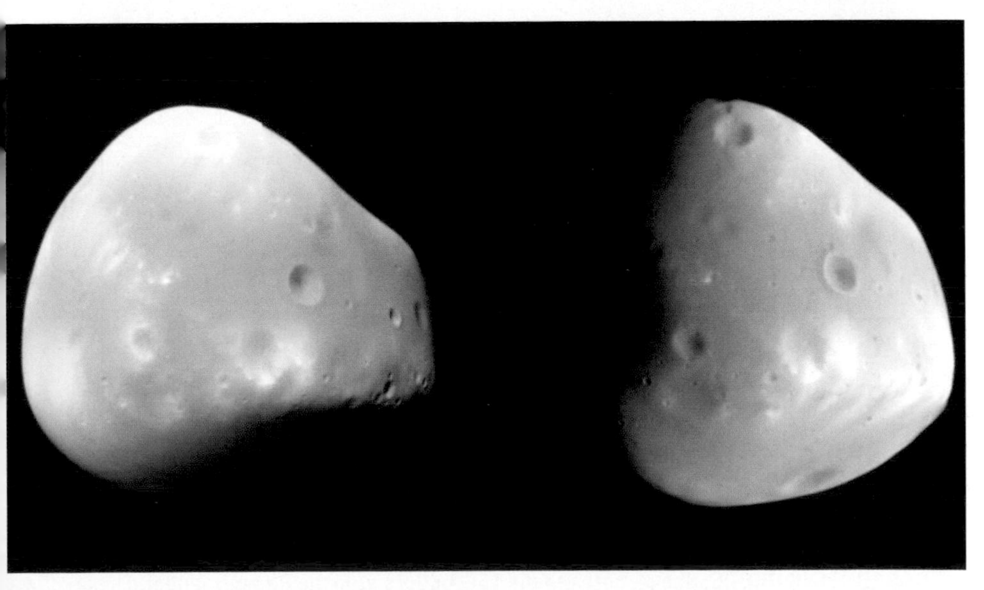

DEIMOS, AHOY: The smaller of the two Martian moons as seen by MRO. At only about eight miles across, it is a tiny body with negligible gravity. Deimos is composed of materials more similar to an asteroid than to Mars; the moon may have been captured in its distant orbit as it drifted too close to Mars. *Courtesy of NASA/JPL.*

DOWN AND SAFE: This artist's impression shows the Phoenix lander with its sampling arm partially deployed. The craft landed in the northern polar region of Mars in May 2008, following the unsuccessful attempt of the Mars Polar Lander a few years before. *Courtesy of NASA/JPL/the University of Arizona.*

DAUNTING DOORS: Shown here are the TEGA doors atop the Phoenix lander. These covers for the sample containers gave controllers much trouble as they attempted to deposit sticky soil samples into the chemical labs within. The doors to the right are ajar; these doors gave controllers ongoing headaches during the mission. *Courtesy of NASA/JPL/the University of Arizona.*

WHAT LIES UNDERNEATH: This image, taken by the camera at the tip of the sampler arm, shows the soil below the Phoenix lander. The white area is ice exposed by the rocket engines during descent. After much puzzling over whether or not ice would be found at the landing site, this area, though inaccessible to the sampler arm, gave mission planners hope that they had found what they came for. *Courtesy of NASA/JPL/the University of Arizona.*

THE BIGGEST YET: The Mars Science Laboratory (MSL) is seen in final preparations. The capsule, larger than the Apollo moonship, encloses the automobile-sized MSL rover, planned to arrive at Mars in 2012. *Courtesy of NASA/JPL.*

THE FUTURE: This artist's impression by NASA artist Pat Rawlings depicts a future Martian explorer arriving at the landing site of Viking 2 in Utopia Planitia. *Courtesy of NASA/JPL*

simile of the rover out to California's Mojave desert and simulating driving on the rock-strewn Martian plains. Like MER, FIDO moved slowly (less that 1 mph) and sported an onboard hazard-avoidance navigation scheme. The investigative tools were similar to Spirit's as well. As the program evolved, the rover was controlled from JPL just as it would be once on Mars. Commands were relayed via satellite with a built-in time delay to better simulate the Martian mission. In fact, the JPL personnel were not even advised as to where the rover had been dropped off, to better maintain the illusion of being on Mars.

The importance of simulating this mission was obvious. MER was a quantum leap beyond Pathfinder. The new rovers were much more autonomous, and indeed needed to be, for they would be covering up to three hundred feet in a day, which was farther than the Pathfinder rover ever got from home base during its entire mission. And with a planned mission duration of ninety days (and much more was hoped for), it was critical to gain experience with the machine.

While there were some basic differences between MER and FIDO—size, weight, and physical operations among them—they were essentially very similar and lessons learned would be of great value. Getting stuck in a crevasse or tipping over due to an overly ambitious climb were a lot less expensive, not to mention final, on Earth than on Mars. On Earth, someone can wander over and kick the rover (and at times, someone did). On Mars, all one can do is invoke harsh language from afar.

The designers of the simulation had done their best to compress twenty days of Mars operations down to about ten days on Earth. This meant focusing on the most important operations and letting a few others slide. Like the grueling Apollo and Shuttle simulations of previous eras, planners deliberately inserted malfunctions and small emergencies into the program to make sure that the rover drivers back at JPL were on their toes.

The FIDO rover covered only about 450 feet during the ten

days of the simulation, and the longest single traverse was about ninety feet. But it was enough: great pains had been taken to include problems and obstacles that would test the controllers' abilities and nerves. When asked, most of them felt as if they had already survived a compressed version of the mission.

As Spirit prepared to begin its historic travels, one programming team was sweating out the upcoming events more than the others. These were the authors of the autonomous hazard-avoidance software, essentially an advanced cruise control with the ability to avoid ramming into things or getting stuck, or so they hoped.

Beyond the driving-and-navigation software, programming was also used to make sure that the mechanical arm housing much of the instrumentation on the rover would not bang up against rocks and dunes. This required the rover to load 3-D images from its mast-mounted cameras into the computer, where it would build a 3-D map of the area being examined. This would then be compared to the projected path of the mechanical arm. If there was a conflict, changes to the arm's path (or suggestions to change the position of the rover) would be enacted to prevent damage to the sensitive electronics.

Overall, important lessons were learned, implemented in updates to software or procedures, then tried again. It's hard to say how many mishaps were avoided by the use of these simulations, but most would agree that they were worth many, many times their cost.

Back on Mars, it was January 15, 2004, and Spirit rolled onto Martian soil for the first time. Like a timid child heading off for a first day at school, the first image Spirit sent down was a (nervous?) look back at the now-empty landing stage. Its own tracks in the ruddy soil led to the bottom of the frame. Mission control erupted in cheers. In just over a minute, Spirit had moved ten feet, much faster than Sojourner. As a sign of how long things take when dealing with the lag time between Earth and Mars, including relay time from the assisting orbiter, the elapsed time between the sent command and reception of the confirmation from MER was over ninety minutes. You wouldn't want to drive in rush-hour traffic this way, autonomous software or not.

The first target, to be reached in about four days, was a rock called Adirondack. It was a low-lying football-sized chunk with flat sides that were fairly smooth. This made it a fine test candidate for the RAT, which could clean and, if needed, grind down the surface of the rock for closer examination. The rock and nearby dirt were examined, and almost immediately the surprises began. The first and most profound was the discovery of a mineral called *olivine*. This is a mineral that is easily altered by water, and the fact that it was here in relatively unaltered form meant that there had perhaps not been as much water in this area over the millennia as had been thought (and hoped for).

Then, just as things were getting interesting, Spirit went quiet. On January 21, eighteen days after its arrival, Spirit lapsed into silence. After a long and knuckle-biting day back home, the robot sent a message to its masters on Earth. It said, in effect, "I AM HERE. I RECEIVED YOUR LAST MESSAGE. I AM IN FAULT MODE—STANDBY." For the next few days, technicians at JPL worked overtime trying to determine if this was a software or a hardware problem (one was probably recoverable, the other probably not, respectively). Then a few more short and cryptic messages came in from Spirit. Frustrated, JPL ordered the spacecraft to downlink an engineering data dump. It eventually did so, and it the problem revealed itself: the rover was having insomnia; it was dropping out of "sleep mode" and using excess battery power in the dark of the Martian night. This was also causing excess heating, which could be a danger to the craft.

Various theories were forwarded by increasingly desperate programmers and engineers. A leading candidate was that the machine was stuck in a "reboot loop," an endless condition that occurs when the computer onboard the rover thinks there is a problem, restarts the computer, only to run across the problem when the computer is booting. It senses the fault and restarts again. The cycle can run forever, or at least until the rover's power supply or circuitry fails. It is a sort of cybernetic Möbius strip of the mind. Spirit had acquired a case of digital obsessive-compulsive disorder.

Programmers came up with a solution and it was sent up. The trick was to avoid a certain part of Spirit's cybernetic mind: the flash memory. It was not unlike a built-in version of the flash sticks we all use today. There was simply too much of the wrong kind of data in this area, and it was causing the faulty reboot sequence.

It took until the thirty-third day since landing, February 6, to resolve the issue, but resolve it they did, and Spirit regained consciousness. The historic mission of exploration was on once again.

Spirit was still parked near Adirondack; it was a bit like someone falling asleep in his soup and waking up, unfazed, to continue eating. Unsurprisingly, little had changed since it went comatose two weeks previous, and the rover resumed its chores as if nothing had happened. The RAT was brought to bear and began grinding away at the surface of the small rock. Controllers were careful—this was the first use of the tool on a Martian rock, and nobody wanted to be the person responsible for breaking it—but despite their caution, a nicely ground disk soon appeared on Adirondack, and the microscopic imager and spectroscopes on the robotic arm were brought into play. While the hole produced was only about one-tenth of an inch deep across an area just under two inches wide, it was enough. The results of this first-ever in-depth investigation of a Martian rock was like the first performance of a well-rehearsed ballet, and with the software issue resolved, things appeared to be back on track, and handsomely so.

The next challenge for the recently reawakened rover was to think on its own. A rock named White Boat was selected as its next target, and this time Spirit was told where to drive, but not specifically how to get there. It would have to use its hazard-avoidance cameras and software to plot a course and execute it. To do this, the stereo camera atop the mast on the rover provided images that the onboard computer used to build a 3-D map of the area ahead. Based on this, the rover would select the safest path, then continually update it with information gleaned from this and the hazard-avoidance cameras. It was a bit like handing your teen the keys to the car for the first time, with the attendant parental nervousness.

Meanwhile, there was more excitement afoot. On the other side of Mars, in a scene not witnessed since the twin Viking landings of 1976, another fireball appeared in the skies above a region called Meridiani Planum. The area had been selected, like Spirit's landing area, from a carefully parsed list of candidates. Of particular interest here was the presence of the mineral hematite, also known in some variations as magnetite, a form of iron that normally occurs in the presence of water. And since possible life, past or present, is the holy grail of Mars exploration, Opportunity, like its twin, would follow the water.

About six minutes after entering the thin Martian atmosphere, Opportunity had bounced to a halt, safe and sound, on Mars. But this rover, Opportunity, had been a bit luckier in its somewhat random final destination. JPL could pick the region, but the final resting place was at the whim of its Superball®-like arrival. For while mission planners had aimed for Meridiani, they had not known Opportunity would end up almost fourteen miles from the anticipated landing zone in an area soon known as Eagle Crater. While the rover could have later driven to such a crater, landing in one was a lucky stroke, and JPL considered the shot a "hole in one."

You see, craters are like holes punched into the crust of a planet to reveal the materials inside. Eagle Crater was no exception, and in addition to what the scientists might have expected to see, they also spotted layers of rock outcrops not far from the rover, about twenty-five feet away. Closer examination revealed apparent sedimentation . . . a possible sign of water-created processes. The layers, ranging from thick to thin (perhaps half an inch to paper thin), were presumed to be either a result of water-borne sediments or deposits of falling volcanic ash. Either pointed to a living, geologically active planet. And one possibility—water—was yet another indicator of the possibility of life. Meridiani was going to be an interesting place.

In early March there was another ripple of excitement through the MER team. Spirit had stopped at another rock, this one affectionately named Humphrey. It was about two feet tall and wide and appeared to be worth a closer examination. Spirit closed with Humphrey and got to grinding with the RAT. Once scientists got a look at the freshly revealed surface, there was a surprise in store. What they found literally sparkled in the sunlight: crystals. If found on Earth, this would assuredly indicate that water had moved through this volcanically formed rock at one time. The same could be true on Mars.

For anyone left from the Mariner 4 years, the scene would be somewhat surreal: a wheeled, mobile, drivable machine with semiautonomous computers onboard was parked next to a Martian rock, grinding away with a rock drill. It then employed devices of which folks from the 1960s could only have dreamed, at close range. The future had arrived.

Opportunity then moved toward its most attractive immediate target: the freshly named Opportunity Ledge. In an area dubbed El Capitan, an outcrop that displayed what were apparently different kinds of layering and weathering from top to bottom. This is the kind of thing that sets a planetary scientist's blood aboil, a four-inch molehill named after a mountain in Texas. After much visual examination, the RAT set to work grinding away. Once this was complete, besides showing the expected profile of a layered rock, two round shapes were visible. Known as "spherules" in the trade, they excited the researchers to even greater heights. About the size of BBs, averaging about one-sixteenth of an inch, these were later named "blueberries." They can be formed by different geological processes, and, as is so often the case in geology, the context in which they are found can have a lot to do with how they are interpreted. But finding them here, inside a rock on Mars, was downright dreamy. More were found nearby, both inside the outcrop and on the ground.

But there was more. Empty areas, called *vugs*, were laced throughout the ground area. These hollows were consistent with comparable terrestrial rocks where something had been dissolved—by water—after the rock formed. To add to the mounting evidence, the spherules were found to be composed of water-spawned hematite. Further work with the spectrometer revealed another mineral, jarosite, which is also formed in water—more evidence of a watery past. This is the kind of evidence planetary geologists love—three distinct sources of aquatic evidence had emerged in one location, and things were looking good.

On March 2, JPL announced that water had at one time flowed through the rocks, changing their texture and chemistry. While these words may not sound profound to a layman's ears, from ever-cautious planetary scientists, they are almost ironclad. There had been water, and probably a lot of it, in Meridiani Planum.

It was the first victory dance, albeit a low-key one. Time to move on.

Spirit meanwhile reached a crater called Bonneville. It peered cautiously over the edge, looking for points of interest. Across the crater, about five hundred feet away, the rover spied its own heat shield, which came to rest there after being discarded during the craft's descent. While the crater looked inviting, and while the end of Spirit's primary mission was drawing near, it was elected not to risk descending into the crater, and Spirit, prompted by its earthly controllers, moved on. Far off, across the floor of Gusev, the Columbia Hills beckoned. Soon the rover was over one thousand feet away from its lander; a pittance by human exploration standards but a vast distance for a Martian rover.

Back at Meridiani, Opportunity had left the rocky outcrop that had consumed so much of its short life. Greater adventures lay ahead, as it spent some time evaluating the skies above—including observation of a somewhat rare transit of the Martian moon Deimos across the face of the small and distant sun. At only nine miles across, Deimos would not block the sun; it was merely a speck on the face of our star. The rover next began a long look at a rock euphemistically named Last Chance, which it had come upon as it neared the end of its primary mission. Then there was a bit of a showstopper.

In a wide-angle picture of the area, a bizarre-looking object appeared. It was the wrong color to be here; it was also the wrong shape. It looked like . . . the head of a rabbit. Now, planetary researchers are not the type of folk given to flights of fancy, and most anything other than a beer can or an abandoned Chevy Corvette® found on Mars is going to find a bunch of men and women huddled around monitors, seeking a logical explanation for what they are seeing. But when something appears that triggers the human impulse to make a familiar shape out of the unfamiliar, especially in a place where it is so unexpected, let's just say that it is a reason to pause and reflect.

And then, just as they were taking a better look . . . the two-inch yellow oddity vanished.

The first assumption was that it must be man-made, something that came from Opportunity as it was landing. Lots of events occur during a landing cycle, and some are somewhat violent. Little bits can fall off a lander as it plummets to Mars, especially when it lands the way the MER spacecraft did, bouncing to a final stop. But the color didn't quite match anything they could think of on the lander, though it had a general resemblance to the material from which the landing bags were made.

A painstaking examination of other images taken earlier

revealed the bunny closer to the lander. Aha. Then one bright soul performed a spectrographic analysis of the object and compared it to the stuff from which the airbags were made—and it was a match. Later the object was spotted hiding underneath one of the petals of the lander. The final conclusion reached was that it was a piece of torn airbag material after all, that had been blown around by light local breezes. Mystery solved, and not too soon, for on a mission of this magnitude, one can't have researchers spending too much time on any one problem unless it is promising of a relevant discovery . . . and while the bunny did appear to be shy in its choice of a final hiding place, it was pretty clear that it was not a living thing nor an ancient Martian artifact. In the end, one of the more arcane mysteries since the Face on Mars had, again, been solved via patient thought and analysis. But it had been fun.

More rocks lay nearby, begging for analysis. Opportunity continued to look at Last Chance, snapping 114 microscopic images of the rock. As an example of what it takes to conduct such a seemingly simple task, controllers had to send up about four hundred commands and reposition the robotic arm two hundred times. But what they found was worth the trouble.

The rock was a sedimentary feature and showed what geologists call "cross-lamination," a sure indicator of water-deposited sediments. Further, these were laid down by *flowing* water, not merely a pool. Evidence was mounting that Opportunity was not merely in an area "drenched in water," as one press release quoted, but better still, had come to rest in a former coastal area of a large sea.

This area was a bonanza for the geologists. Opportunity went on to examine more rocks and features before leaving Eagle Crater. One such area was called Berry Bowl, because it was host to a number of the "blueberries" seen near and inside of El Capitan. The hematite spherules were more far common than suspected. Although the individual "blueberries" were too small for the spectrometer to be entirely accurate, here there were a sufficient number of them that the instrument could get a reasonably good reading; and they checked out as hematite.

Opportunity soldiered on, wheels slipping a bit in the loose soil as it departed Eagle Crater. It was off to explore the promising Endurance Crater. At about 430 feet across, Endurance was far larger than Eagle, and other than a lone rock in the area, the only feature of interest nearby. It was 2,300 feet away, much farther

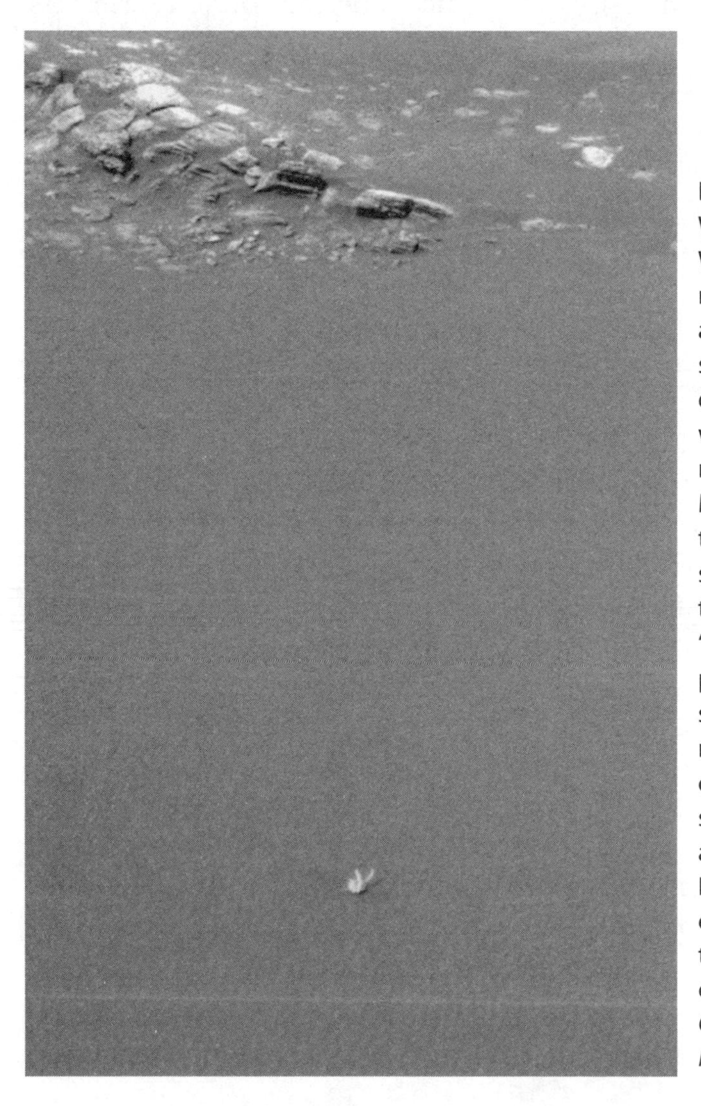

Figure 19.1. WASKALLY WABBIT: While at no time did anyone in mission control doubt that this was some man-made part of the MER spacecraft, this oddly shaped apparition, dubbed "the Rabbit," did play hide-and-seek with the rover team and create a brief surge of intense activity on the Internet. In the end, it was identified as a piece of the airbags. *Courtesy of NASA/JPL.*

than the rover's travels to date. The Sunday drive would be leisurely and slow, but not without more potential for discovery.

Across the planet, Spirit continued its mission, steady and sure. If rovers had emotions, surely this machine—only the second mission to land on Mars and first of its generation—would resent its higher-profile sibling. By no fault of its own, Spirit had landed in an old lava flow inside a crater, with little to offer beyond some smaller craters that might provide some insight. Compared to Opportunity, which had struck pay dirt, it was a distant second at this point in time. A bit like hitting a vein of pyrite when your neighbor hits gold. Not the worst thing, but a pale second place.

Fortunately, robots have no hearts to break. Spirit looked toward the Columbia Hills, just over a mile distant, and rolled onward.

CHAPTER 21

DR. STEVE SQUYRES AND THE MARS EXPLORATION ROVERS

DREAMS OF ICE AND SAND

Steve Squyres came to space exploration with a pedigree: first there was Cornell, one of America's finest universities, where he earned a doctorate while studying under Carl Sagan. Then there was that postdoc at NASA Ames in California, not given lightly. *Then* Cornell asked him to return to his alma mater as a professor, something not often done in the top tiers of academia. Oh, and let's not forget the H. C. Urey Prize from the American Astronomical Society in 1987, or his 2007 Benjamin Franklin Medal in Earth and environmental science.

But to talk to him, it is clear that all this pales next to his time spent on Mars, at least through the telerobotic senses of his Mars Exploration Rovers. His interest in exploration of faraway places started early: "I grew up in a household where people were interested in science; I was interested in science at a very young age. I was also very interested at a young age in exploration. I loved climbing mountains when I was a kid; we used to take faraway vacations like Colorado. I was also as a child fascinated by the history of exploration, so I used to read a lot of exploration books about the Arctic and that sort of thing. Science and exploration were all things I was interested in."[1]

Studying geology seemed to be a gateway to far-off places on Earth, and for many geologists, it is. But there was a problem: there were not very many places on the maps of Earth that were still terra incognita.

"What I discovered after two or three years of being a geology major was that the scientists who have been doing geology on Earth for centuries have [already] done a pretty good job at figuring things out. So for someone who thinks of himself from an explorer's perspective, it felt like filling in the details, and I was interested in more of a blank canvas."

That canvas would be his, and filled with a ruddy red . . .

"My third year at Cornell I took a course on results of the Viking mission to Mars, which was happening at the time, 1976 and 1977; it was a graduate-level course, so because of that we were expected to do a piece of original research for our term paper. I got a key to the Mars room, which stored pictures from the Viking expedition and orbiters. I remember going into that room and looking at the pictures for four hours, trying to find out what I was going to write about, but I walked out thinking about what I was going to do with myself. And then it was obvious, *this* is what I was looking for.

"I switched gears, went right into planetary science, applied to a bunch of different grad schools. This was at the time of the Voyager mission, and while I was deciding where I was going to go to grad school, Carl Sagan approached me and told me he'd be interested in having a grad student on the Voyager mission, so I worked on Voyager while I was in graduate school. It was a fabulous experience and I've been doing planetary exploration ever since."

Since Voyager, and in addition to professorial duties, Squyres worked on both the Magellan Venus project and the Near Earth Asteroid Rendezvous mission with JPL. But Mars was still out there, beckoning. He got interested in the idea of Martian rovers as a follow-on to Viking, and soon found himself the principal investigator for the Mars Exploration Rovers.

"The primary goal of MER was to go to two places on the Martian surface, trying to learn what conditions were like there in the past, and then find out when they might have been habitable. Mars is a cold, dry, desolate place to stay, but . . . in the past, conditions may have been different. So we tried to choose two places that appeared from orbit, first and foremost, [as a good] place to land on, but also, places that appear [to have had] traces of water, and to try to really read the story in the rocks, and to see what conditions existed and how habitable it really might have been."

About the time JPL restarted Martian exploration after a lengthy post-Viking hiatus, a picture began to emerge that the Mars Exploration Rovers would be able to augment in a spectacular fashion.

"According to previous missions, there has been abundant water on the Martian surface in the past. Now, evidence of water is not entirely credited to rovers. If you look at Mariner 9, you can see the first evidence of water channels and so forth, so we noted in the early 1970s that there was once water on Mars. But what we have been able to do is to fill in a lot of details regarding the chemistry, the kinds of processes that were going on there. This a very rough number, but for the first billion-or-so years of [Mars's] history, conditions on the surface were warmer and wetter, there was alteration going on, there were hydrothermal systems; it was a much different place than it is today."

By 2004, Squyres had two well-equipped rovers on the Martian surface, one on either side of the planet. One of the first discoveries, and not a heartening one, was the presence of large amounts of a rock that he wished had not been there: olivine.

"It is a mineral that tends to be present in unaltered igneous rocks [that is, unaltered by water], so finding it was a disappointment, because that was one of the first things that made us realize that we landed on a lava flow instead of . . . on sedimentary rocks. We landed on a lava flow . . . one at least a billion years old." And lava flows, while interesting, are much less promising for this type of exploration than something indicative of watery activity.

"It took awhile for it to kind of sink in what we were dealing with, that the sediments we were looking for were completely buried in the lava. Once we finally realized that, we had to move somewhere else, and we decided to head for the Columbia Hills."

This was a geologically interesting region nearby that had been named in honor of the astronauts lost when the space shuttle *Columbia* broke up during reentry in 2003.

The Columbia Hills were over two kilometers from Spirit's landing site, and given the slow rate that the rovers were allowed to move, it was a long haul. An astronaut on Mars wouldn't think twice about covering the distance; controllers driving an aging rover, which had completed its primary assignment but seemed to have plenty of life left, had to consider their options with care.

"The rover was designed to last for ninety days, and drive over 600 kilometers [about 375 miles] in its lifetime. On day 100 of our ninety-day mission, we decided that we needed to go somewhere different, and the Columbia Hills were 2.5 kilometers [about 1.5 miles] away. We had already voided the warranty of the vehicle, and we had no idea how long it was going to last, so we just went as fast as we could. But we got there. We reached the hills on sol 156 and spent the rest of the mission there."

Meanwhile, across Mars, Squyres had another rover to worry about. Opportunity had enjoyed the good fortune of landing in Eagle Crater, which provided a whole different set of options.

"The biggest discovery from Opportunity came within the first 60 sols, so we lucked out. We discovered a giant impact crater that basically had all the things that we could have wanted exposed in the wall of the crater. In two months, all the most important science was revealed to us. Since then, we've been taking advantage of the fact that Meridiani Planum is very smooth, very flat, very capable for driving, to cover a lot of ground. Our strategy is to go from impact crater to impact crater. We're driving around on layered sedimentary rocks that were horizontally layered, which meant that you're driving along a flat surface, and you're basically seeing the same rocks over and over again. So what you need is some capability to get down below the surface. Now, we didn't bring a drill rig, but Mother Nature has dug many craters for us on Mars, so we had to go to those craters and then find a big one, and go into it to explore below the surface. We [now have] something like thirty-one kilometers [about nineteen miles] on the odometer so far.

"We're heading towards Endeavor Crater, and it is huge, [a] twenty-plus-kilometer-diameter feature [about 12.5 miles across] formed a very long time ago, before the sediments were even deposited. Pieces of the rim of the crater are sticking through the sedimentary layers, and those pieces of the outer rim are older than anything we've ever seen before. So we're closing in on the crater, all the never-before-seen materials, with an old, tired, beat-up rover."

Spirit has ceased to operate, but in a way it lives on within the mission of Opportunity. Soon the second rover will reach its destination, very possibly its last, and provide us with new answers and new questions.

It's been a long haul for Squyres. There were times he wasn't sure the MER mission would happen at all: "One of the high points was, simply, the launch. I mean that very seriously. I went through ten years of writing unsuccessful proposals to NASA before they finally said yes. And then, we were flat-out canceled and brought back to life three times! When we finally were given the go-ahead to do the mission, we [had] thirty-four months to do a job that would properly take about forty-eight, so at times it just seemed inconceivable that we would ever get to the launch pad. Just getting to Florida was a miracle; just getting a shot was a high point.

"Then there was the day when the first rover drove off the lander and had six wheels in the dirt and we were ready to really explore. That moment was really the culmination of seventeen years of work that we had done to make that happen, and after that it was just our rovers and what Mars gave us and how well we could perform with what we had. It's just been a seven-year-long adventure with one discovery after another. To me, because of all the years of effort that was put into just making the thing happen at all, those events were the ones that affected me the most deeply."

Squyres falls silent after this for a moment, and perhaps it is because he realizes that it not only deeply affected him, but the rest of humanity as well.

Opportunity's mission continues.

CHAPTER 22

MARS IN HD

MARS RECONNAISSANCE ORBITER

It was as if everything from before had been viewed on VHS tape and the newest images were in Blu-ray®. The images from JPL's latest, the Mars Reconnaissance Orbiter (MRO), were *that* good. This mission was a whole new ball game in orbital imaging.

For one thing, the cameras onboard were the best to date. The other scientific instruments accompanying them were also cutting-edge. But it was the radio dish that was perhaps the most arresting feature of the orbiting spacecraft. While the craft itself was about 1.5 times larger in size than earlier orbiters, the dish was almost as wide as the spacecraft and at least twice as wide as previous transmitters. At almost ten feet across, it was a huge jump in bandwidth. Combine this with the latest in data compression, and it was like going from dial-up to broadband in one fell swoop. In fact, by 2008, after just about 1.5 years of operation, the craft had sent home over *fifty terabits* of information, or more than *all* of JPL's planetary missions to date. That's an impressive accomplishment in anyone's book, and a bigger "pipe" by an order of ten over earlier craft.[1]

Like other recent Mars orbital probes, MRO was built for JPL/NASA by Lockheed Martin. It was large by Martian orbiter standards and about the weight of a small car (2,300 pounds dry). In a nod to modern materials engineering, much of the structure was made up of carbon composites in addition to the more traditional aluminum and titanium. It was also the first spacecraft designed from the ground up specifically for the stresses unique to aerobraking. The computer brain of the craft was somewhat behind the cutting edge for off-the-shelf hardware of the time, a radiation-hardened, military-grade version of Motorola's Power PC® chip (the same unit that drove older Macintosh® G3 computers). The use of a commercially retired chip (the Mac was onto an advanced G5 by then) was not due to budget constraints or lack of foresight; rather, it reflected the need for a robust, proven, and evolved version of what would do the job. By planetary-exploration standards, this was still cutting-edge. Reliability and the ability to handle intense radiation from solar flares was key (thankfully, the military had already proven the chip's ability to handle radiation from nuclear attack, so space-borne radiation should be child's play). Few such chips are guaranteed to withstand the rigors of interplanetary space travel.

The eyes and ears of the spacecraft involved improvements on existing technology and some new designs as well. The High Resolution Imaging Science Experiment (HIRISE) was a new generation of camera and lens technology that would look down on the planet from on high. Previous cameras had boasted of being able to see things in visible light as small as about twelve feet across; the new camera would resolve items as small as three feet. The camera was also able to image in near infrared (much like Odyssey), which would allow it to see things not visible in normal light, such as dirt-covered gullies, channels, and other water-created artifacts. Mars was getting ever closer.

No Mars mission would be complete without a spectrometer. The Compact Reconnaissance Imaging Spectrometer for Mars (CRISM) device sought out minerals that, as usual, extended the search for a watery past for Mars. Visible landforms were one thing; CRISM would search for the invisible indicators of features from the past, now buried or covered, such as dried pools, former thermal vents, dead hot springs, or dry lakes.

A device called the Context Imager (CTX) would create lower-resolution images to augment those of CRISM and HIRISE. The idea was to take wider-angle pictures of the exact locales observed by the other higher-magnification devices to provide a context (hence the name) for those observations. An example might be if one of the other devices showed evidence of water-deposited sediments such as a streambed or dried pond. The CTX camera would snap a wider-scale image (about twenty-five miles wide) that might show the former pond to be a part of a wider feature created by watery sediments or volcanic activity. It was a simple and elegant way of collecting data concurrent to the more powerful instruments. In addition, both CTX and HIRISE would be able to produce stereo (3-D) images of the surface below, continuing a trend of stereo imaging begun with the Viking probes.

Yet another device to take visible pictures was the Mars Color Imager (MARCI). This camera's job was simple: create regularly updated images of the broader planet for use in building seasonal climate and weather maps of Mars over time. It was a bit like a multispectrum stop-action camera trained on a busy street in Hollywood, providing a parade of otherwise pedestrian images that, over time, would create a fascinating picture of changes and trends on Mars.

The Mars Climate Sounder (MCS) was a modified camera that was able to sense in both visible and infrared. Its job was to measure, in deep "slices" of the atmosphere below, a profile that included temperature, humidity, and dust content, in high to low altitudes. Over time this would build up a more complete story about Martian weather in great detail.

A Shallow Subsurface Radar (SHARAD), supplied by Italy, probed for more water on Mars, as did so many other devices. What made this one remarkable was its reach: it could look almost three thousand feet deep! Where previous instruments had been able to see a yard or so into the dirt, rocks, and soil below, SHARAD promised to reveal much more complete results, albeit at less fidelity (features less than about four hundred feet wide would not be seen).

A gravity-field instrument would allow MRO to measure the Martian gravitational field in some detail, something not taken for granted since the Apollo program detected large variations in lunar gravity fields during the moon missions. These variations, dubbed "mascons" (mass concentrations) threatened to throw off orbital calculations and became a prime concern during that program. Future planetary missions would include devices to measure such things, no longer assuming that planets would have a nice, even gravitational field such as Earth's.

Finally, a device lyrically named Electra (for once, a proper name and not an acronym) was a radio that would act as a Martian GPS for the Mars rovers and also relay information from them and future landers to Earth. It would also be able to help track newer spacecraft on their way to Mars from Earth.

It is worth noting that a craft such as MRO has very critical navigational and positioning needs. With such high-magnification instruments aboard, small errors would be rapidly compounded. One of the quiet heroes of such a mission was the collection of devices that would orient and point the craft to specific spots for imaging and data collection. No less than sixteen sun sensors were affixed, along with two advanced star trackers (which had come a long, long way since the Canopus star tracker of Mariner 4). Additionally, two inertial measurement units, not dissimilar from the technology that allows an iPhone® to sense its orientation, were included to provide precise measurements of velocity in any direction, augmented by laser-enhanced gyroscopes. If nothing else, MRO would know *exactly* where it was at all times.

To best utilize this data, twenty small rocket motors were onboard, along with six larger thrusters for gross changes in speed. These tiny thrusters would allow for incredibly precise positioning of the spacecraft, and were further augmented by a system of flywheels called *reaction wheels*, used to stabilize the probe during operations.

MRO left Earth in 2005 for the now-familiar trip to Mars.

Slinging into Martian space too fast for immediate proper orbit, as was now traditional for Mars probes launching on smaller, affordable rockets, it began aerobraking operations in March 2006. After the successes of the Mars Global Surveyor and Mars Odyssey, you might think that aerobraking was old hat at the lab. But not so, for, as it turns out, JPL had been planning these maneuvers under some mistaken assumptions about atmospheric density and had been somewhat lucky on previous missions.

Let's review. When a probe like MRO reaches Mars on a trajectory that requires aerobraking, it still fires the same slowing rockets that a mission like Viking, which did not require aerobraking, did. But it has less fuel than Viking and is therefore unable to slow as much, so it cannot force itself into a nice, round orbit. It enters an elongated orbit, shaped like a large egg. In MRO's case, aerobraking saved over 1,300 pounds of fuel—a huge amount when you are launching things from Earth.

The orbit the probe entered upon reaching Martian space was lopsided, with the farthest point being twenty-eight thousand miles from Mars. The nearest point during aerobraking was about sixty miles. The final orbit, the lowest yet attempted (which would allow for close-in use of the new high-powered cameras) would be 196 miles and circular.

To accomplish this aerobraking maneuver, one must plan very precisely. And to do that, you must know all the variables, such as the atmospheric density of Mars, exactly. Otherwise, you will not slow the craft sufficiently as it dips into the friction of the atmosphere, or you will slow it too much or perhaps even damage it. So JPL uses a mathematical model to calculate this value, which had worked well in the past. But MRO was to be a lower, more carefully positioned orbit, and extreme care would have to be taken to trim the aerobraking orbit just right.

When dealing with lower maneuvers in the atmosphere, one must take into account that the polar regions are very cold, resulting in denser air there. Another reason that density can vary from day to

Figure 21.1. AEROBRAKING: The MRO spacecraft is seen in this artist's impression during aerobraking maneuvers. It took about five months to circularize the spacecraft's orbit. *Courtesy of NASA/JPL.*

day (even at the same altitude) is the lack of oceans on Mars. On Earth, the oceans are able to store large quantities of heat during the day and then release them slowly at night. On Mars, with no liquid water, no such balancing mechanism exists. Temperatures can fluctuate quickly and dramatically, by well over 100°F per day. Although the atmospheric pressure is just a tiny fraction of Earth's, these large temperature fluctuations affect the air density very quickly.

And there was yet another reason that the planning for these maneuvers was critical and was keeping controllers up at night: the space around Mars was getting crowded. There is the Mars Odyssey probe, Mars Global Surveyor, the European Mars Express, and a host of older orbiters slinging around the planet. During aerobraking, when you are changing altitude continually, you must plan to miss these other craft, as mission managers tend to get very cranky if *your* spacecraft smashes into *their* spacecraft during an orbital maneuver. Of the twenty-six rocket firings to alter the aerobraking maneuvers, six were designed to avoid colliding with other probes. It was a very complex piece of mathematical wizardry to pull off from many, many millions of miles away.

Nonetheless, five months later, MRO had reached a stable, circular orbit almost two hundred miles above Mars and began its primary mission. The probe began investigations intended to send home about five thousand images yearly. The primary mission was intended to be two years, then optional mission extensions would be considered. Ultimately, as with Mars Odyssey, researchers will have to split time between their own projects and large amounts of data being relayed home from the Mars Science Laboratory rover. MRO is already in extended-mission mode, probably one of many.

On to the science.

Since all Mars missions have a goal, one way or the other, of searching for water, MRO did its part. With all the experience gained over the years, the complex of instruments the designers had clustered on the spacecraft—the most powerful to date—worked in brilliant harmony to provide answers. And the results have been rewarding.

One mystery of Mars has been to decipher the distant past of the northern regions. The southern hemisphere of Mars is heavily cratered and represents an older, less disturbed surface. But the northern regions, with their enormous volcanoes, had at some time been flooded with volcanic eruptions; lava had flowed, hiding much of the ancient surface there. And with that went much of the possible evidence of the watery past researchers had been seeking. But from its low-altitude vantage point, MRO was able to point its high-resolution cameras at these regions, seeking minute geological detail, and send back amazingly defined images of the terrain. And though it has thoroughly mapped only about 1 percent of the planet to date, the finds began accumulating quickly.

But, as always, there were issues. About two months after the science operations began, MRO experienced difficulty with the Mars Climate Sounder instrument. A so-called stepping mechanism malfunctioned, and the aim of the instrument became slightly off-axis. By December the problem had not corrected itself, and regular use of the device was abandoned. A partial work-around was later devised, and the MCS was pressed back into use, but not without compromises.

Additionally, the image-sensing electronics in the HIRISE camera, the CCDs (much like those in your home video camera) began to lose individual pixels and some electronic noise was found in the incoming images. While this was not a deal breaker, it didn't make anyone on Earth happy. Again, a work-around (utilizing a longer warm-up period for the electronics) minimized the problem, but it still remains as a weak point for the telescopic imager.

But perhaps the most alarming issue presented itself in mid-2009 when the onboard computer began resetting itself—shades of Pathfinder and Spirit. Perhaps the Great Galactic Ghoul has adapted itself to the modern era and figured out how to fool spacecraft computers into thinking that they are faulty, triggering a shutdown and restart. In any event, software changes seem to have alleviated the problem for the present. But the computer team remains vigilant. It took over two months to eventually solve the problem.

These issues did not stop the mission, however, and discovery after discovery continued to stream in. Wide measurements of the northern ice cap revealed more water than could have been hoped for, almost two hundred thousand cubic miles (not square miles, but cubic miles!) of water ice. This is equivalent to almost a third of Greenland's ice sheet and accounts for a lot of the "lost" water on Mars.

Some of the evidence of a wet past had come from the observation of phyllosilicates, heavily hydrated minerals formed in water.

But the lava flows in the north had covered any evidence of these. The researchers needed a hole punched in this lava layer, and through extensive observation with their new orbiting toy, they got one. Nine craters in the lava fields were investigated, and each of the craters revealed hydrated minerals in the older layers below.

More craters were targeted, and a number of them showed bright blue-white materials on the surrounding ground. A few passes of the orbiter later, the material was slowly disappearing. From the rate of evaporation and the colors seen, it was clear that, once again, water ice had been found (the color indicated that it was almost 100 percent pure water ice). Just to be sure, the spectrometer was pressed into service, and, sure enough, the spectral signature matched that of water. To make it even more dramatic, some of these were about forty-five degrees south of the pole. By now, spotting ice in the polar regions of Mars, and as far south as sixty degrees, which is roughly equivalent to Anchorage, was not a shock. But finding it at the latitudes equivalent to Paris or Seattle was. It just kept getting better and better.

All this talk of water may cause the untrained ear to become a bit jaded. So there is water on Mars . . . big deal. But it may be just that—a very big deal. How this ice is formed so close to the surface is still a bit of a mystery. Mars's atmosphere is far too thin to support liquid water on the surface, and the formation mechanism for these ice patches is still not well understood. One theory involves a process that on Earth is called *frost heave*, in which small amounts of water can remain liquid around a grain of solid ice, even at temperatures below which it should freeze. Pressure causes this liquid water to migrate upward, where it then freezes, forming a lens-shaped structure on top of the soil below. Why this is important (beyond the pure geological implications) is that this process, which keeps water temporarily liquid in certain places near the surface, could form environments where bacteriological organisms could thrive. And as biological studies on Earth continue to find basic life-forms colonizing areas as diverse as undersea hot vents and the frozen dry valleys of Antarctica, the idea of water-bearing areas on Mars is tantalizing.

Continuing the search for water, in 2009 MRO used its camera cluster and spectrometer to image the vast reaches of Valles Marineris, the huge, hemisphere-girdling canyon that straddles Mars. In a region named Noctis Labyrinthus, scientists were looking for light-toned deposits (LTDs in the vernacular) indicative of water activity. They examined ten LTDs, which turned out to be troughs in the canyon, and found things they did not expect to see. The instruments on MRO, working in well-planned harmony, identified clays, hydrated silicas, and sulfates—all of which pointed to yet more watery activity sometime in the past. Some of the formations were dozens of miles across.

Small differences assumed large proportions. An example was one trough where most of the water-affected minerals were buried under later, wind-driven soil, but some was visible in the upper walls of the trough: a sure sign that the water-affected area was older than the trough itself. Another featured (water-derived) clays buried beneath newer plains. These and other findings, while seemingly innocuous, indicate a confused and jumbled timeline of multiple "wettings" of the area, multiple water-inundating events, which is in itself a major discovery. In short, Mars had not had just one watery time, but a number of them. This bodes well for a complex geological history, and again offers a possibility for the existence of past life.

Just how the water arrived in these troughs was not evident. It has been hypothesized that it may have been melted ice from the volcanoes nearby, or some subterranean hydrothermal event. Since then, clays (representing, again, water processes) have been identified throughout Martian bedrock, so it is not a unique occurrence.

These discoveries resulted from the elegant and coordinated use of the context camera and the higher-resolution HIRISE imager. First the low-magnification CTX would spot something interesting, then the HIRISE imager would zero-in with its telescopic lenses and take a hi-res picture for detailed investigation. Finally, the spectrometer—the CRISM instrument—would take a careful look at the area and provide its chemical analysis. In this way, working in perfect three-part harmony, MRO was able to first identify, then analyze in detail, these kinds of soil deposits in high-resolution. The deposits dated back somewhere between 3.5 and 1.8 billion years ago—hardly recent, but important nonetheless. The troughs themselves seemed to have developed somewhere in the middle.

Glaciers were discovered in regions that first sparked interest among researchers in the days of the Viking orbiters; they surrounded the edges of cliffs in Martian valleys. They are lobe-shaped and gently sloping, and many are covered with debris and soil. Again, these are areas that apparently store huge volumes of water ice.

MRO also provided some of the first looks at earthly artifacts as well. When imaging Victoria Crater, the HIRISE camera captured an image of the plucky Opportunity as it went about its long traverse of the edge of the crater, making out the body of the rover and even the shadow of the camera mast. The HIRISE was also able to later photograph the Mars Phoenix Lander as it slowly descended toward the Martian north polar region in May 2008, dangling from its parachute. This was particularly exciting due to the fact that the landing event is short-lived and was snapped from an oblique angle—not the camera's strongest mode of operation. Both images were evocative and, besides being strong technological accomplishments, highlighted the infinitesimal human footprint on the Red Planet.

The images of Victoria Crater provided another benefit—to assist MER mission planners in their guidance of Opportunity into Victoria Crater. These were all examples of multiple Mars missions working together, something never before accomplished on such a scale. And not only did they provide data from different

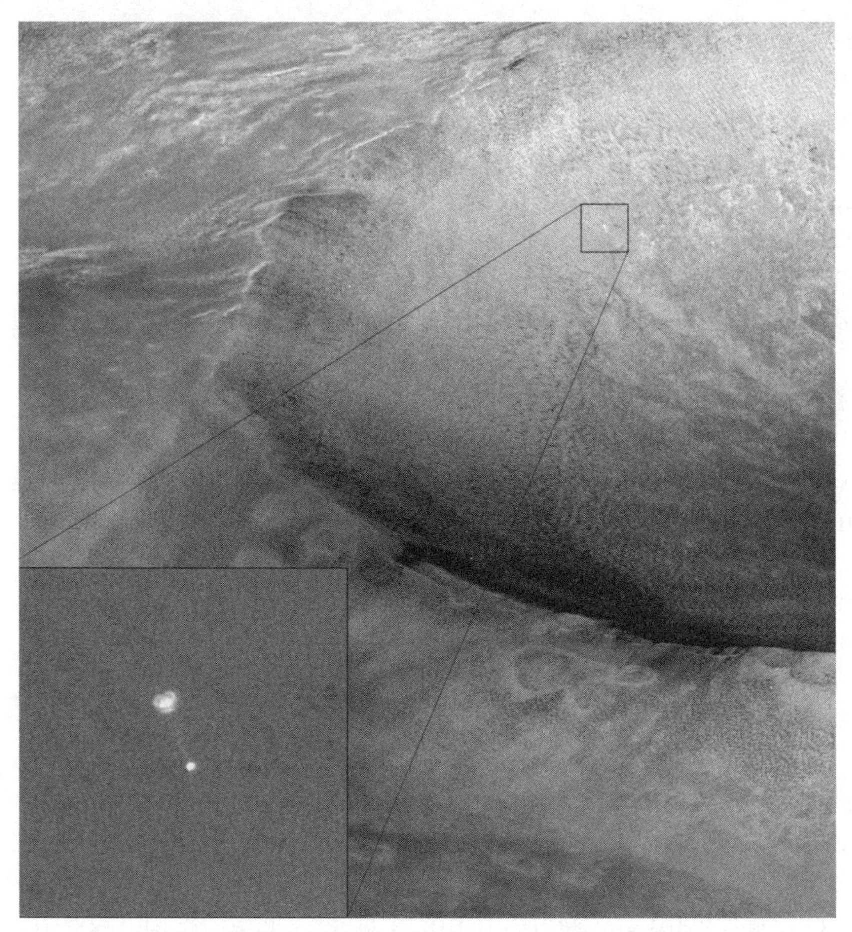

Figure 21.2. FAMILY PORTRAIT: The Mars Reconnaissance Orbiter was able to image the Phoenix lander during its parachuting descent. This was a first in unmanned space exploration. *Courtesy of NASA/JPL/the University of Arizona.*

perspectives; these multiple machines were actually able to help each other. Mars exploration was becoming a team sport.

There is one more area of discovery to be attributed to MRO: the observation of "real-time" or near-real-time events. One such occurrence was an avalanche seen by HIRISE in February 2008. A sloping hillside in Mars's northern regions had a huge dust cloud billowing from the base of the slope, clearly demonstrating a huge movement of material to the base of the incline—an avalanche. The camera was actually targeting a dune field nearby as a part of its regular survey of carbon dioxide frost. The avalanche was a serendipitous capture, only noticed later when a mission scientist was reviewing the images. It may have been caused by water ice being exposed, suddenly sublimating, or evaporating, and undermining the soil.

Another new feature was spotted via a comparison of images taken by MRO and Mars Odyssey. Again, it was a team effort. MRO's wide-angle context camera had snapped an image in 2008 in which a small black dot was spotted. No big deal. But when compared to an image from Mars Odyssey from 2006, there was no such spot. Excited researchers aimed MRO's HIRISE camera at the same place and found a small, otherwise insignificant crater about eighteen feet across. The hole itself was not what was spotted by MRO—it was darker material exposed by the shock of the impact, which blew the overlying, lighter surface away. They had spotted a brand-new crater. While not hugely significant, it was a fun moment . . . and there is always room in science for some fun.

The Mars Reconnaissance Orbiter has, to date, returned more data than all the missions launched past the moon. It is a huge amount of scientific material, and it has been estimated that if just the HIRISE images were shown consecutively for ten seconds each, it would take over four years to view them. So make sure you are well-stocked for snacks.

And MRO isn't done yet. Stay tuned.

CHAPTER 23

DR. RICHARD ZUREK, MRO

I CAN SEE CLEARLY NOW . . .

From his home near the foothills of Southern California, Richard Zurek can look toward JPL and see the object of his study, some forty years since he began to study it: the air. When not enjoying the view, he hikes the local mountains. But his consuming interest is, of course, Mars. And he has been a part of seeing Mars as it has never been seen before, through the high-fidelity eyes of the Mars Reconnaissance Orbiter. His inspiration to study Mars would sound familiar to many of his compatriots at JPL . . . the flight of Mariner 4.

"I was in high school at the time, the Mariner 4 flyby. Now, as recently just a year or two before Mariner 4 was launched, you [could] still read papers on the dark areas, what did they represent, was it vegetation, what was this planet like? And at [that] time they overestimated the amount of atmosphere that it had. . . . [They estimated] the atmosphere to be ten times thicker than it really was.

"Mariner 4 was the first to see that this thing was about 1 percent of the Earth surface pressure. . . . [T]hen [there was] the challenge of orbiting a spacecraft around the planet; Mariner 9 succeeded in doing that, becoming the first earthly object to orbit a planet other than the Earth itself. So first we have Mariner 4 and say, 'maybe there's nothing here,' then the flybys for 6 and 7 and the controversy on whether or not they detected chlorophyll, and it turned out to be another weak indication of carbon dioxide, so arguments about Mars—what is it like, what it's not—continued."[1]

Mars was proving to be more adept at hiding its secrets than anyone would have guessed, especially after the dramatic results of Mariner 4. Mars transformed from a planet populated (in the minds of some) by an advanced civilization of water-hoarding civil engineers to a dry desert wasteland overnight. But then the follow-up flights made it clear that things were not so simple. The highlight of this may have been when Mariner 9 arrived at Mars, only to see a huge planet-girdling dust storm. But sticking out of that storm were the peaks of three huge volcanoes on the Tharsis Bulge. Mars was revealing its past slowly, making us work for it.

"This happened to us a couple times, certainly with Mariners 4, 6, and 7, and then Mariner 9 changed the idea again, both in the sense of 'yes, there were mountains, and there were vast channels on the planet,' and 'yes, the atmosphere was thin, but it was vigorous enough to be able to sustain regional dust storms into a global haze that could blanket the entire planet,' and so Mars sort of came alive again."

Inspired by JPL's missions of Mars exploration, Zurek decided to aim for the stars: "I went to the University of Washington, and mathematics graduate school, but I was pretty certain as I soon as I got there that I didn't want to be a theoretical mathematician. I chose the University of Washington because I knew they had a department of atmosphere and science that took people with math and physics backgrounds. I had that, [so I] transferred into that department, and it just happened that they had a new professor. This new professor had an interest similar to what I'd had as a kid, so I was fortunate. Not only was he a new professor at the school, he had an unfilled assistant role that he needed to fill, and so I walked in and I was a good match.

"When you get to a mission like MRO, which came along after the great discoveries of Mariner 9 and the Viking orbiters and landers, things kind of build up incrementally. But [the results were] profound, and I think a couple of the things really impressed me. The ground is so oddly patterned. There were polygons, and fractures; it just looked like a surface that has had many places that were once wet and had since dried out. It's like looking at mud-flats, only the scales are bigger than that, and I think that tells us two things. First, there's been these drying out episodes, that there's still ice in the surface of the planet, and the other one is of course the composition measurements that have indicated that you have these areas where these minerals are present. When you put that whole picture together, what you see is that what you're looking [at] is an ancient surface that's been covered up.

"With the ancient surface, we were looking at a lot of water interaction in it, because it altered the composition of the surface; actually changed the material into sulfates, and carbonates. Seeing [this] early history, and trying to figure out just how widespread it was is interesting. [However] the fact [that] there are different minerals indicating different kinds of watery environments, some of them more acidic than others, to me, once again increases the potential that Mars might have developed life somewhere."

Every mission has its highs and lows. MRO was no exception, and coming on the heels of the twin failures of Mars Climate Orbiter and Mars Polar Lander, the stakes were still high, even years later: "Just getting into [Martian] orbit, getting safely into low-altitude orbit, was a high point. Scientifically, seeing these constant changes of Mars climate from an ancient period [when] water was active, there must have been water surging through the ground, and in salt lakes, to form these mineral deposits that we see today. And then you have the more recent climate change . . . in the polar caps, and the buried ice deposits. It's all very interesting. Of course, as an atmospheric scientist, dust storms are still a big question. Why do some storms get huge, [involving a] large fraction of the atmosphere, and at other times we only have [a small one] out of three Mars years?" These and other questions were nagging Zurek when the MRO entered its science orbit, bringing its revolutionary imaging capability to Mars.

"That's what resolution does for you. When you're looking at a lower resolution, you don't see the variations that are there; they are fuzzed out by the inability to resolve. [At] the scale in which we're now seeing the planet, the colors [can be] stretched so we can see the variations. We do that because it tells us different things, the bluer materials are often the sand-covered things, the white-zoned areas are often things that get altered by being in contact with water."

The resulting images are truly magical and have a beauty all their own. MRO brought a new dimension to the visuals coming back from the Red Planet: "The principal investigator of the high-res camera says that one of his prized moments was when he was here at JPL, and he was looking for an office. Someone told to him to go to the end of the hall where the abstract painting was, and to turn left. So he walked down the hall, and the abstract art turned out to be his camera's picture on the wall. The beauty of seeing these different elements at high resolutions and seeing that landscape . . . shows that they've done a great job."

But MRO's job went beyond basic science. It was pressed into service as a relay for the Mars Exploration Rovers, and also to help identify landing areas for the upcoming Mars Science Laboratory (MSL).

"Today they announced the landing site for MSL, it's going to be Gale Crater, which was one of the four finalists. I took great pleasure in that, because our MRO team provided the basic data by which precise selections were made, both [in] terms of the engineering safety and also in terms what interesting things are in these places.

"So, we expect a lot more in MRO. It has enough fuel to go for another decade. [The spacecraft is] working so beautifully, sending so much data back, we're looking forward to continuing for hopefully nine more years. Things are looking pretty good at the moment."

Here's to another decade of Mars Reconnaissance Orbiter, and the secrets that it will reveal.

CHAPTER 24

TWINS OF MARS

SPIRIT AND OPPORTUNITY, PART 2

On sol 159, the indefatigable rover Spirit reached its first stop at the Columbia Hills. It had been dawdling, pursuing its scientific endeavors, at Lahonten Crater since sol 118, driving around the rim of the two-hundred-foot depression. Then the long drive to the hills began.

At the base of the hills, it spent a full twenty-three sols studying a feature known as Hank's Hollow, and in particular an odd-looking rock named Pot of Gold (there was no lack of imagination within the MER team). Lo and behold, there be hematite in these hills—Spirit was catching up with the discoveries of its precocious twin, Opportunity. The rock was described as looking "as if somebody took a potato and stuck toothpicks in it, then put jelly beans on the ends of the toothpicks"; a colorful way to describe the oddly shaped, softball-sized rock. The process of formation was thought to be water based, especially because hematite is usually found in the presence of water. But there are chemical processes that could result in such a feature, so it was studied exhaustively.

After investigating with its "sniffer," the rover repositioned itself (which took several sols, due to the slippery and treacherous nature of the surrounding soil) and ground away at another side of the rock with the RAT. Whatever formed this rock took a long time and was not a mild process. It was harshly eroded and knobby, and the nodules on the rock appeared to be cousins to the "blueberries" found by Opportunity, half a planet away.

Spirit left Pot of Gold to investigate more strange-looking rocks nearby. Dubbed Cobra Heads, these rocks, as did Pot of Gold, appeared to have come down from higher up in the Columbia Hills. The Cobra Heads were probably the cores of more normal-looking rocks that were left behind when the softer surrounding layers wore away to leave the harder interior behind, silently hissing at Spirit.

At this point, while still very capable, Spirit's spirit was being dampened by some problems. The mechanical arm was suffering from a bit of dysfunction in one of the motorized joints. The radio was having issues, making the reception of commands sent from Earth difficult. And the front right wheel was draining far more electrical current than the other five, and this drain was increasing. It was an indication that the gearing that drove that wheel was failing, probably due to sand, dust, and grit having crept inside. To make matters worse, the season was advancing and Spirit was receiving less sunlight on its solar panels every day.

Still, the rover continued its labors, taking a northerly course along the base of the hills, investigating as many rocks as it could along the way. Then, on sol 239, the rover powered down for solar conjunction, when Mars is on the far side of the sun from the Earth and out of communication with JPL.

When we last left Opportunity, it was embarking upon the long and lonely journey to Endurance Crater. On April 30, 2004, the mammoth feature loomed into view. About 430 feet across, Endurance Crater is named after the ship captained by Sir Ernest Shackleton on his epic voyage to Antarctica. It was a fitting name for a landmark so prominent in Opportunity's life.

The rover paused at the lip of the crater, taking its time to survey the sixty-foot-deep interior. As with most impact craters, Endurance ripped through the surface of the planet, exposing the strata below. Finding a suitable point of entry, Opportunity, shaking its metal arm in the Great Galactic Ghoul's odious face, took the plunge and began the risky descent into the crater. The point of entry was named Karatepe ("Black Hill" in Turkish), and the rover took a tentative drive inside the crater rim, then stopped and backed out, just to make sure that it could. It was a bit like watching Neil Armstrong on that July night in 1969, coming down the ladder of the Lunar Module, then before setting foot on the moon, hoisting himself back up the rungs just to be sure that *he* could.

Nobody wanted Endurance Crater to be the final resting place of Opportunity, though everyone involved knew that it might be. As it turned out, the rover spent the next *six months* driving around Endurance Crater. The first stop was a patch of exposed bedrock strata. Opportunity got up close and personal with the outcrop and sampled each layer of the strata as carefully as possible. It noted that both the texture and the chemistry of the rocks varied with depth, with the lower layers being the oldest, just as on Earth. And both magnesium and sulfur declined in abundance as the rover's sniffer went from the upper (younger) to the lower (older) rocks, once again confirming the presence of (and alteration by) water in the past.

Moving on, more "blueberries" were seen in the rocks and scattered about the crater itself. Soon the same solar conjunction that idled Spirit forced Opportunity to cease operations for two weeks as the two rovers waited out the silence from Earth.

As Mars moved back within range of the ground controllers, Opportunity moved to a rock unlike anything seen in the Meridiani Planum area. Dubbed Wopmay (after a semifamous Canadian bush pilot), its lumpy, serrated surface promised more signs of alteration by water. More "blueberries" were found here as well, and the sightings of these was becoming almost boring . . . almost. Again, signs of a watery past.

About this time you might well be wondering just how much evidence of water in Mars's past a planetary scientist might need. It's a good question. But with such an arid planet meeting virtually every spacecraft sent there, and the promise of life, past or present, dangling on this finding, there could be no such thing as too much evidence. Not yet anyway.

Tellingly, the investigation of this rock corresponded with the publication in a scientific journal of the results of the MER missions' efforts to find evidence of a watery past on Mars. The paper included the efforts of 122 authors and was confidently published in *Science*, perhaps the most important journal appropriate for this kind of news. This made it official: there has been large amounts of water on the Red Planet, and the evidence would just keep piling up. Besides the "blueberries" and various chemical residues, there were cavities, the oddly named vugs, in many of the rocks investigated that were indicative of crystals that had been dissolved over time by water as the rocks lay immersed. This finding, wondrous in the extreme, told of not just large amounts of H_2O, but also that it had existed as a liquid on the surface of the planet for a long time in the past. And this indicated a thicker, warmer atmosphere way back when.

If there was a downside, it was that this water would have been salty and acidic and not overly friendly to life. But life finds a way, as it has in so many hostile environments on our own planet, so this was far from a deal breaker.

Returning to Spirit and its numbing slog across the Gusev region, the rover had found something interesting (well, it was *all* interesting, but this was a major find). The mineral in question was goethite, kin to the jarosite found earlier by Opportunity, and it was another sure indicator of water sometime in the past. The beauty of this mineral is that, unlike hematite, which *usually* forms in a watery environment, goethite forms *only* within such an environment. Anyone sweating out the declarations in the *Science* article could take a deep, relieved breath.

Making spirits brighter, to coin a phrase, the worrisome friction buildup in the front right wheel was diminishing. Nobody could be sure why without a mechanic's house call to Mars, but the fact that controllers had been "babying" that wheel may have helped. All they knew for sure was that it was drawing less current, and that was a good sign of mechanical health.

Spirit now ascended a formation known as Husband Hill (named after the NASA astronaut who died in the shuttle *Columbia*'s demise) in an effort to ascertain how high the water might have once stood. The hope was that it was not merely an underground store but might have pooled on the surface for some length of time, as it appeared to have done at the Opportunity site. Confirmation was the name of the game.

Then Opportunity, never content to allow Spirit to bask in the limelight for long, found something wonderful.

Ascending out of Endurance Crater, the rover had spied its own heat shield from its fiery descent onto Mars. Scientists wanted to examine the structure to see how it had been affected by its hot journey through the Martian atmosphere, and the results were interesting, if not remarkable. Nearby was an interesting rock, dubbed, appropriately enough, Heat Shield Rock. As the rover neared this stone, it began to look eerily familiar to some on the ground, and very unlike the others they had been investigating. Soon, it was confirmed. Opportunity had found the very first meteor found on another planet. Where it came from (besides space) and when it arrived would remain a mystery, but finding it was excitement enough. And it was an iron-rich meteorite, not the more common "stoney"-type. Of course, meteorites are found on Earth, so the rock itself was not the real news. What it would reveal about the area on which it sat was the interesting part. It could help to answer the question about whether or not Meridiani Planum was gradually eroding away as so much of Mars is, or if it was still being built up in geological terms.

Soon, Opportunity moved on. The region it was crossing was sufficiently flat that it allowed controllers to test more autonomous driving than before, depending on the rover's onboard computer to avoid hazards and obstacles. So the first part of each drive was based on a course set by ground controllers,

Figure 23.1. THE HEAT SHIELD: Upon leaving Endurance Crater, Opportunity spied its heat shield, jettisoned during its descent from space about 330 Earth days before. *Courtesy of NASA/JPL.*

then switched over to the onboard navigation system. This was a good test of JPL's software, and the rover passed with flying colors. Up to a quarter mile was covered in this mode, a new record.

Months later, in June 2005, Opportunity had a near-death experience. It had gotten bogged down in a small sand dune, and it struggled for almost five weeks before freeing itself. Had the arm been able, it might have wiped a bit of sweat from its metallic brow, but Opportunity instead took a time-out to access its health before carrying on. At the same time, JPL wanted to know why Opportunity had gotten stuck in an area that looked so much like regions the rover had crossed with ease before.

Far away, in December 2005, Spirit spent some time staying up at night to observe a meteor shower. Mars, like Earth, passes through swarms of cometary debris every year. The comet in question was the famed Halley's Comet, and the resulting show did not disappoint. It was what researchers call "bonus science," an unplanned addition to the rovers' busy schedules.

By mid-2006, Spirit had to find a place to spend the harsh Martian winter once again. The fact that both machines had performed far beyond expectations, and had vastly exceeded their warranties, was remarkable enough. But nobody wanted to lose one to the winter cold for lack of planning.

After descending from Husband Hill and investigating a formation called Home Plate, Spirit parked itself on a north-facing slope to allow for maximum exposure of its solar panels through the winter months. Spirit would need all the power it could get to keep warm through the long cold period. An added drama was that the front right wheel, the previous source of concern, had finally given up and was no longer functioning. From now on, Spirit would be dragging one foot as it ambled across Mars.

Opportunity had just wrapped up four months at a 1,000-foot wide crater called Erebus, after the volcano in the Antarctic.[1] It had found thin, rippled rock indicative of flowing water—another important discovery. Sitting water was turning out to not be unique in Mars's ancient past, but evidence for wind-blown water was still being sought to confirm interactions between weather and standing water.

Soon Opportunity continued on its trek, heading off on a long drive to the largest and most promising crater yet, the half-mile-wide Victoria Crater, named after one of Ferdinand Magellan's ships. Over two hundred feet deep, Victoria had exposed bedrock layers over 130 feet high, and promised a treasure trove of new information. The larger and deeper the crater, the more of Mars's geological past would be revealed.

Opportunity was near the equator, while the ever-unlucky Spirit was in the southern hemisphere, so while Spirit shivered through the winter, Opportunity was still receiving power from the sun. She was ready to tackle her biggest challenge yet.

In October, Opportunity began driving the rim of the vast crater. About this time, the Mars Reconnaissance Orbiter began work overhead, and was able not only to spot tiny Opportunity going about its chores, but also to begin to assist in planning its future by looking inside the crater itself. The ultra-hi-res images MRO sent home gave MER planners much-needed data and allowed them to plan with far greater confidence.

From its vantage point, perched on the rim of the crater, Opportunity had provided researchers with sufficiently detailed images to show that the layers of rock in the crater wall were diverse and promising. There were specific divisions, indicative of environmental change. This is just the sort of thing they had come for and, if the rover stayed cooperative, would investigate.

Besides investigating the crater from a safe vantage point on the rim, Opportunity was seeking a safe entrance for a possible drive down to the interior. This was no small feat. Victoria was not an exact circle like many craters; its sides were heavily scalloped. While this offered a variation in incline for the drive inside, it also made navigating the rim trickier. The machine needed the best possible combination of slope and material composition to risk entering the crater. Too much sand and it could slip. Too many rocks and it could be blocked. Nobody wanted a replay of the "rover-stuck-in-a-dune" drama of a few months back. Caution was the watchword.

After a few months of carefully picking its way around the rim of the crater, Opportunity joined its sister rover, Spirit, in a bit of downtime for a software update. This improved version would allow the rovers to better "think" and take more independent action. One example was the ability to perceive changes in visual and cloud patterns, so that rather than slavishly sending back hours of images of the horizon through which researchers would have to sift to find a change indicating, say, a dust devil, the rover would be able to "sense" the differences and send back to Earth the relevant images, dismissing the rest.

Another new capability was called *visual target tracking*. Previously, when the rover was driving in autonomous mode, it could not track the location of a boulder or other impediment as it moved. Once the visual aspect of the item in question changed, it became a new item in the rover's mind. This software update allowed the onboard computer to track the offending item as the rover moved—vital for ongoing mission success.

Finally, a new routine had been programmed to allow the rovers more autonomy in the use of their robotic arms. Previously, when the rover neared a target for investigation, ground controllers had to assess safe approaches and give specific instructions to the arm. This new software allowed the onboard computer to determine the risks itself and preselect targets, and approaches, for a closer look.

While these may seem mundane changes to some, the ability to cut down on the twenty-minute delay in communications (each way—so double that and add time to go through an orbiting relay such as MRO) allowed ground controllers and the rovers to accomplish much, much more in a shorter time. And with the clock ticking on each of the rover's mechanical and electronic systems, time was critical.

It was now June 2007. Opportunity had spent a few months driving around a prominence of Victoria called Duck Bay. The rover had investigated what it could around the rim, while at the same time seeking a safe route for descent. By the end of June, it returned to the most promising site. The slope here varied from fifteen to twenty degrees and was mostly exposed bedrock, which, it was hoped, would minimize the dangers associated with slipping on sand or dirt.

At this point, Victoria had been monopolizing Opportunity's attentions for almost two and a half years. Things move slowly in Mars rover-time. It was time to make the leap of faith and head toward the crater floor. If tension could be eaten, flight controllers would have their lunch set out for months to come.

Then, bad weather intervened. Massive dust storms kicked up, raging for over two months. To the rovers, the threat was not just of a loss of solar power to their batteries, but also of reduced visibility and windblown grit. But they hunkered down and sat out the storms. It was a long wait. Some days the solar panels could only generate power equivalent to that needed to light a 100-watt bulb for an hour. They could not operate normally in this condition and went into reduced power mode, restricting activities to a minimum. Even so, there was real concern. At one point, only 1 percent of the normal light reaching the rovers in daytime was available, and the batteries were dying. The skies overhead were almost opaque. And if power was too low, the rovers would "safe" themselves by going into a condition called "Low Power Fault." This would put the machine to sleep, from which it would rouse itself briefly each day to check out prevailing conditions and available power. The problem was that nobody was entirely confident that the machine would wake up at all if it slept too long. It was a real nail-biter of a time.

But fate smiled upon the mission, and by September the skies had cleared and it was time to commit. The depleted batteries had recharged and were fit for duty. On the eleventh, Opportunity made its first tentative drive into Victoria. Again, it crossed the threshold, moving about thirteen feet downslope. Then it reversed direction to ascertain how much it might slip on the way out and to make sure that it could retreat to safety if it had to. Two days later, the die was cast, and Opportunity headed toward the crater's interior, its home for the upcoming year.

Within Victoria was a bright band of sedimentary material that girdled the crater, which was inevitably dubbed the "bathtub ring." This turned out to be a water table from a time not long before the impact that created Victoria some millions of years ago. Then, as the rover descended, it began to spot more "blueberries," the hematite spherules seen across Meridiani Planum. But these were far larger than those investigated earlier, again suggesting that the farther back one looked, the wetter Mars got. It also argued for much of this watery interaction to have occurred underground, rather than on the surface. The mysteries just kept unraveling.

Interestingly, on the way to Victoria, the number and size of the "blueberries" sighted had been decreasing steadily. At the same time, the rover had been gaining elevation, with Victoria sitting about one hundred feet higher than Endurance. But once the rover dove into Victoria Crater and lost elevation, the "blueberries" got larger. Again, water processes were suspected.

As Opportunity drove around the interior of Victoria Crater, it explored many of the bedrock outcrops visually. Most were named after capes and bays from the age of exploration on Earth. As the rover was able to gain more sampling results, what had been observed previously at Endurance and Eagle Craters was confirmed and recognized as regional processes, that is, they took place across Meridiani Planum and not just where they were first seen.

While exploring Victoria, Opportunity began experiencing trouble with its robotic arm again. This was a replay of the problems experienced in 2005. The balky motor was taking more current to work than it should, so engineers had to increase the amount of electricity being used, but not to the point where it could cause a larger problem. It was decided now to change the way they treated the arm while the rover was driving. Previously, the mission rule was to stow the arm under a restraining hook while a rover was moving. But the motor giving trouble was needed to unstow. So rather than end up with a trapped arm, the programmers instructed Opportunity to unstow the arm after each day's work rather than spending the night with the arm under the hook. A small change, but a potentially limb-saving one.

Opportunity's adventures within Victoria Crater continued. The final results would take time to compile, but it was clear that water was in the mix in a profound way. It had come and gone over billions of years. And always there was the wind; in fact, it appeared that most of the scalloping was done by the wind. Opportunity had been able to look at vertical strata of rock over thirty feet high; some was clearly the result of sand dunes compacting into strata layers. Also found outside the crater bowl were

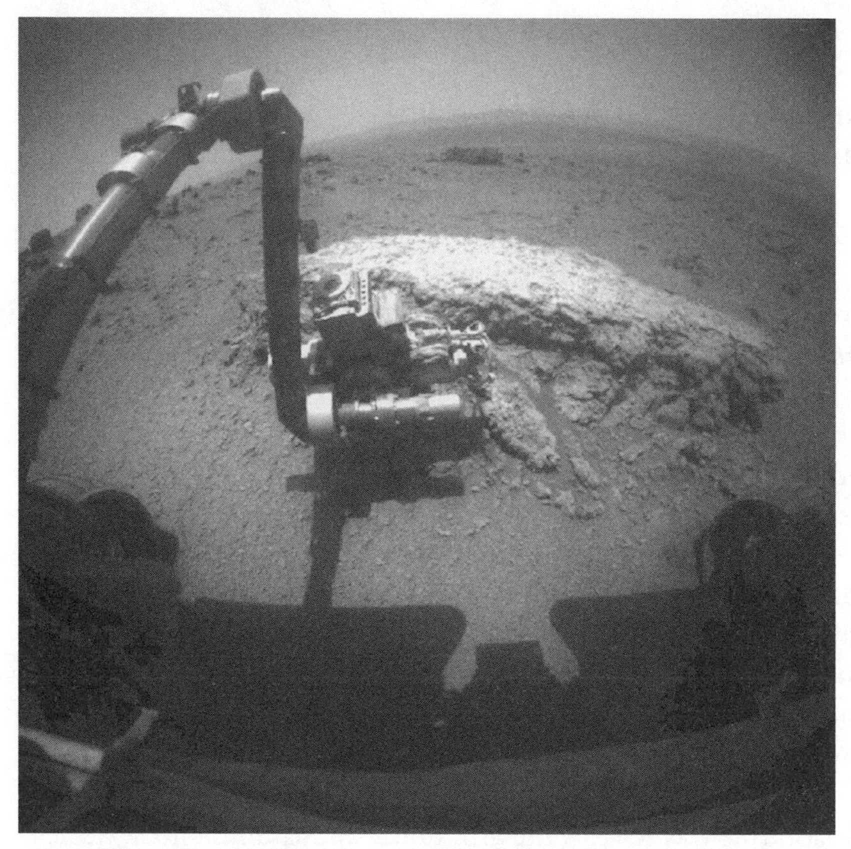

Figure 23.2. ARTHRITIS: By the time it left Victoria Crater, Opportunity began having motor trouble with its manipulator arm. The arm, highly engineered and requiring lots of torque to operate, was critical to the mission, so controllers took to babying it whenever possible. *Courtesy of NASA/JPL.*

bits of meteorite that may have been parts of the body that created the impact crater so long ago.

Other processes and results were noted—most important, the continuing evidence of not just water in the past, but indications of how the water existed as it was affecting the rocks. In this area, the water seemed to have been operating underground more than above, which could allow it to continue to change the rocks long after Mars lost the ability to support water in liquid form on its surface.

While these results were being digested by the investigative teams on Earth, Opportunity made the sometimes-treacherous climb out of Victoria. A year after entering the crater, the rover used the very same tread marks it had left when it entered to chart a safe path out of the depression. At the end of August 2009, the rover completed the one-month drive out of the crater, crossed the sandy barrier at the rim, and was once again on flat ground.

After a brief period of checkouts, it began what would be a multiyear, ten-mile drive to its next objective: the massive Endeavor Crater named after Captain Cook's famous vessel. The journey was expected to take over two years. The adventure in Victoria had consumed almost half of Opportunity's time on Mars. Before departing the area, it investigated more rocks around the edge, targets it had noted but passed by in the drive to get inside Victoria. A further bounty of science resulted, largely confirming what the scientists already knew: Mars was once a wet, wet world.

Endeavor, its next target, was over twenty times larger than Victoria, almost fourteen miles across, and promised an even greater scientific bonanza. As this book goes to press, Opportunity will begin its descent into this larger impact crater to begin a new round of investigations and discoveries for as long as it can manage to do so.

Across the dusty planet, Spirit continued operating, but barely. By late 2008 it had begun showing increased difficulty in operation, and in January of the following year, the onboard computer began to behave erratically. At times it would download driving instructions and not execute them, and at other times it would not fully report the activities of the day. Spirit's health continued to decline; in addition to the front right wheel (which now merely dragged along), its flash memory was becoming faulty. Even so, as recently as January 6, it had stopped to research a rock called Stapledon, which contained high amounts of silica, possibly indicating a former watery hot spring or steam vent.

By April, Spirit's onboard computer was again being testy and

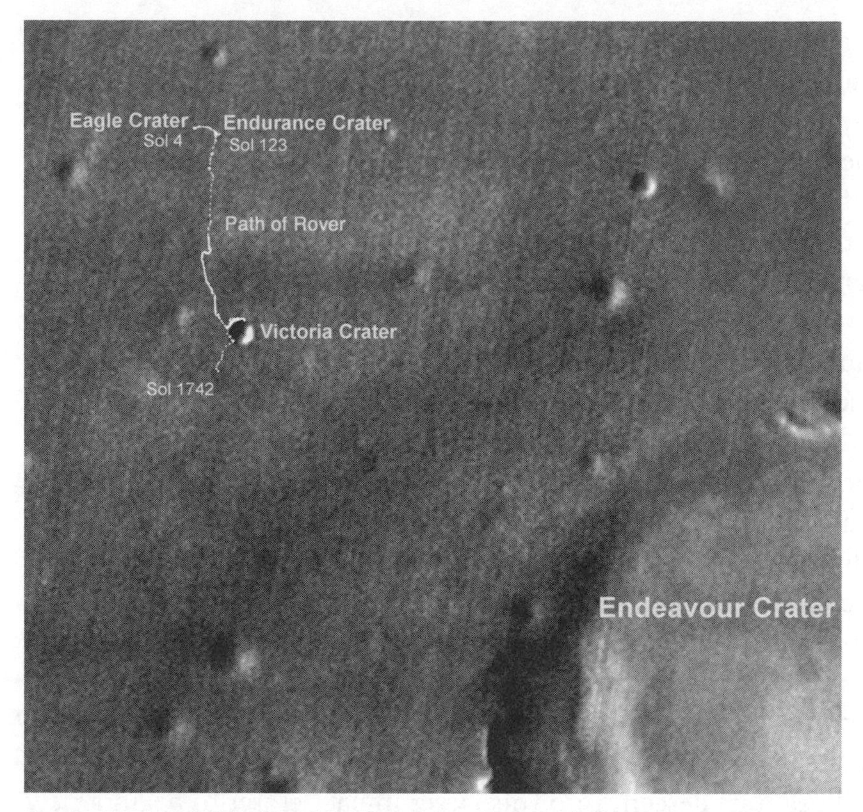

Figure 23.3. ENDEAVOR CRATER: Leaving Victoria behind, Opportunity headed off on the long trek to Endeavor. The scale of the journey, compared to those it had already taken, is seen here. *Courtesy of NASA/JPL.*

began to reboot spontaneously. The flash-memory problems were increasing, resulting in increased amnesia. Then, in June 2009, it became lodged in soft sand; attempts by controllers to unstick the machine were meeting with little success. It had been driving across a crusty surface but broke through into the softer sand below. With winter coming, if the machine could not be moved to another spot that would afford the increasingly dusty solar panels a better angle toward the weakening sun, its future did not look bright.

To make matters more frustrating, it had become clear that the region Spirit had been investigating was not so dull after all. While Opportunity had discovered a formerly very wet and acidic environment in Meridiani Planum, Spirit's home had been found to be a steamy, violent place. More investigation was clearly warranted, but only if the robot could be moved from its current location.

In late March 2010, Spirit stopped talking. The suspicion was that it had entered a hibernation mode, a result of increasingly low energy reserves in its batteries. Already the daily communications with the rover had been cut to once a week in an effort to conserve power. But even that had proved to be too much. As the Mars Odyssey probe wheeled overhead, it listened intently for a signal, but none came.

In late July, controllers began sending commands to Spirit in an attempt to awaken the rover. It was now deemed safe to do so, as the local environment was emerging from the Martian winter and the power drain, if communication occurred at all, would likely not be fatal. But the rover remained silent. It was hypothesized that it may have lost so much internal data that the clock in the onboard computer may not have known what day or time it was. If this had occurred, the computer should reset the clock and begin a sequence of listening for commands for twenty minutes out of every hour of daylight.

By January 2011, Spirit had been off the air for nine months. Still, controllers continued to try to awaken the rover. It was the local Martian spring, and this was the last best hope for communication. Soon solar opposition occurred, and then, once Mars emerged from the temporary blackout, efforts began again. But it was not to be.

On Wednesday, May 25, JPL sent up its last plea for Spirit to phone home. This was greeted with silence. Earth-based radio dishes had double-teamed with the orbiters around Mars in attempts to reestablish communication, but all had been for naught. Spirit, after being stuck in soft sand and enduring the coldest temperatures ever encountered on Mars, had died. After a career of almost seven years on Mars, the rover had succumbed to the harsh elements. Still, having vastly outperformed its initial mission of ninety days, the machine had outdone itself and performed brilliantly.

Opportunity, meanwhile, has completed the drive to its next, and most exciting, target. The rover had covered about fourteen miles since it landed, almost three times as far as Spirit managed. In a fitting epitaph, JPL named Opportunity's point of arrival at Endeavor Crater "Spirit Point." Who says engineers and scientists are not sentimental?

CHAPTER 25

FROM THE ASHES, LIKE A PHOENIX

When one hears the word *Phoenix*, one usually thinks either of a hot city sweltering in the Arizona sun, a Grammy-winning French rock band, or the mythical bird representing rebirth. The last of these is what the planners of Mars Phoenix hoped for when naming the polar lander tasked with succeeding where the Mars Polar Lander mission (MPL) failed when it crashed into the frigid wastes of the Martian north pole.

It was not a perfect analogy; for while Mars Phoenix was made largely of hardware recycled from a previous lander, it was not bits of a twin of Mars Polar Lander, but rather the later (and mothballed) Mars Surveyor 2001, canceled in a period of shock and internal review after the twin failures of the aforementioned Mars Polar Lander and the Mars Climate Orbiter. Fortunately for JPL, the Surveyor had been carefully packed away, and much of the machine was reincarnated in Phoenix.

The mission had origins unlike most interplanetary missions of discovery. After the failures of the earlier lander, various parties within the close community of planetary scientists had been talking, and out of this came a proposal to NASA from the University of Arizona. It would be inexpensive, yet would retrieve the science lost when the MPL crashed. Phoenix would also address the problems that plagued the failed MPL mission. It would be run as a tight ship. It would also be done largely off-site. It was enough to excite and terrify conservative NASA managers at the same moment. And, amazingly, it was ultimately approved as a part of NASA's new and (relatively) inexpensive Scout program.

Of the many improvements made to Phoenix over the failed Mars Polar Lander, the method used to shut down the descent engines was perhaps most important. MPL had used a "shock sensor" that was supposed to have triggered the engine shutdown when the lander set down. Unfortunately, an unrelated shock felt during descent (probably the deployment of the landing legs), while MPL was still over one hundred feet up, shut down the rockets prematurely, and the lander crashed. Phoenix would return to a simpler method, similar to one used as far back as the Apollo lunar lander: switches on the footpads would signal *actual* touchdown. This was one of a number of improvements made in an attempt to ensure a successful mission.

The Mars Phoenix mission was unfamiliar ground for NASA. It would be the first mission in the space agency's history to be led *and operated* by an academic institution, the University of Arizona. This was worrisome to an organization used to complete control.[1] To make things more complex, a number of foreign universities would contribute instruments and expertise to the mission, including institutions in Canada, Germany, the United Kingdom, Finland, Denmark, and Switzerland. The result was a potpourri of scientific and technical input. It is important to note that while the University of Arizona operated the spacecraft, the navigation and landing were controlled by JPL in Pasadena.

The probe landed successfully in May 2008. It was the first lander to travel to the polar regions of any planet. And Phoenix was unusual in other ways. It was small in comparison to any other Mars lander except perhaps Pathfinder, weighing in at just under eight hundred pounds. About five feet across (eighteen feet with solar panels deployed), it was almost the same size as an individual Mars Exploration Rover and would look downright puny next to Viking. And after the bouncy-ball landings of Pathfinder and the Mars Exploration Rovers, Phoenix signaled a return to the rigors of a powered descent, where rockets must handle the final moments of a soft, pinpoint landing. It was not a mission for the faint of heart, especially with the fifteen-minute delay from Earth to Mars for communication. At times, add an hour or two for the orbiting Mars-probe relay, and you have a real issues. Planning was key.

As a largely recycled spacecraft, Phoenix was a compromise mission. It was designed to be a low-cost attempt at a Mars lander (in this case, less than $500 million). This had a number of real impacts on the mission design. And the stakes were high, as this was one of the first planetary missions since the twin losses of Mars Climate Orbiter and Mars Polar Lander; JPL could not afford another failure. Perhaps for this reason, as much as any other, the design and control of the mission followed an unusual path.

While built by Lockheed Martin, Phoenix was designed by a consortium of academic and NASA/JPL personnel. The mission proposal originated from the University of Arizona, and parts of the spacecraft (notably the camera) were actually built there. An ad hoc mission-control center was also sited at the campus, looking more like a low-end software company than a deep-space mission-control room. Many of the staffing needs were met by hiring young and inexpensive talent; some were even fulfilled via the use of grad students. Not surprisingly, many of the key players were from the Mars Pathfinder mission, itself a departure from traditional JPL methods. This was not your father's Mars mission.

Even the software intended to run it was recycled from programming originally designed to operate a Mars *orbiter*. It had to be rewritten and repurposed for a lander, resulting in some last-minute sweat and angst as the software team tested and retested the command structure to make sure it would meet the demands of short-term surface operations. Much of it was left open-ended; changes and new instructions would be regularly uploaded from the University of Arizona controllers and JPL via the Deep Space Network. This recycling of existing software created many sleepless nights for the coding team.

As with previous missions, the processing power of the flight computer was not particularly robust by consumer standards. In fact, the flash memory was only one hundred megabytes; far below off-the-shelf flash drives that were offering over four gigabytes. But given the expense of flight-rated, radiation-hardened components, it was what the budget could support. As with many of JPL's spacecraft, the CPU was the venerable RAD 6000 chip manufactured by IBM, which had been extensively proven in spaceflight, having powered the computers of the Mars Exploration Rovers, Mars Pathfinder, and Mars Odyssey, among others. This was a proven design, and there were many available engineers and programmers who knew how to squeeze the maximum performance out of the chip. It ran at a nonblistering thirty-three megahertz and cost upward of $250,000. It was a far cry from the two-gigahertz Mac G5 available down the street for under $2,000.[2]

The instrumentation on the Phoenix lander was a compromise of light weight, low cost, and high reliability. Deciding what to include was, as always, a study in creative compromise.

Paramount for a lander of this type and purpose was a robotic arm. Unlike Viking's spring-steel "wind-up" arm, it utilized a more traditional and simple hinged "elbow." Designed to extend almost eight feet from the lander, it was capable of digging about eighteen inches deep. Besides a scoop, there was a drill-like rotating rasp for shaving ice samples (a true drill would be too heavy). A small camera was affixed to the end of the arm. As powered landings were still rare at the time (as opposed to the more passive beach-ball approach) another camera, the descent imager (complete with a microphone) was installed to transmit video of the landing. Unfortunately, at the last minute an electronic flaw prevented the use of the camera or microphone. Important data and a wonderful PR opportunity were lost.

The primary camera, the surface stereo imager (SSI), was similar in design to that on the top of Pathfinder's rover, Sojourner—largely because it was built by the same team and had worked well on that mission. It would extend vertically on a mast from the lander.

To search for organics (precursors to life) in the soil, a set of small, high-temperature ovens called the Thermal and Evolved Gas Analyzer (TEGA) would use a spectroscope to analyze gasses baked out of soil samples. The TEGA consisted of eight tiny, one-use oven compartments, each about the size of a pen. The instrument would look for water, carbon dioxide, and organic elements such as methane. It shared a more-than-passing resemblance to the ovens found in the Viking lander thirty years previous.[3]

Phoenix was packed with other instrumentation. The Microscopy, Electrochemistry, and Conductivity Analyzer (MECA) device would examine soil particles on a microscopic level, using a so-called wet chemistry lab or WCL. The WCL had test chambers where purified water would be added to soil samples and sensors would measure ionic activity, looking for biological compatibility of the sample (by earthly standards). This would give some idea of how able the soil would be to harbor microbial life. This was an important distinction from the Viking missions: rather than searching for life, this would search for the ability to *support* life.

Also inside the MECA was the Thermal and Electrical Conductivity Probe (TECP). This would measure soil temperature, humidity, thermal (temperature) conductivity, electrical conductivity, and other properties of the soil fed to the MECA. An optical microscope was included to take pictures of these samples at an extreme magnification, using arrays of multicolored LEDs for different results under different colors of illumination. Each color would reveal different properties in the magnified samples. The final component of MECA was an atomic force microscope (AFM) that shared space with the optical microscope. Small silicon crystal tips would brush across the sample, measuring repulsion from the sampled soil to analyze its composition at the atomic scale.

Atop Phoenix, a meteorological station would provide ground-level measurements of the Martian day and night. The technology used ranged from a simple telltale (not unlike a wind sock used at airports) to a highly innovative and complex LIDAR (laser-powered radar) that measured dust, ice particles, and moisture in the air. Of course, temperature, humidity, and air pressure were measured as well.

As always with a lander, the decision where to set down was a long and arduous process. The vastly improved images available from the Mars Reconnaissance Orbiter and Mars Odyssey helped a lot; rocks as small as twenty inches across could be seen now. This was a far cry from even the Pathfinder mission, when the landing zones were selected by a combination of medium-resolution imagery, intuition, and luck. And, to bolster confidence further, Phoenix had fourteen inches of ground clearance, as opposed to Viking's 8.5 inches. Nevertheless, the final selection would not be made until after Phoenix had left the launch pad.

Phoenix set down in an area called Green Valley in Vastitas Borealis ("Northern Waste"), not far from the north Martian pole, on May 25, 2008. It was the first successful powered landing since Viking 2 in 1976. This region was finally selected for smoothness and other safety considerations, and it was also where the largest concentration of water ice (besides the pole itself) had been found to date.

Among its first duties were to transmit data about its orientation and status back to JPL. As it turned out, things had gone as well as they could have dreamed; the lander was just about level. Next, the giant solar panels were unfolded to give the hungry batteries the power they needed. As quickly as possible, team members needed to establish how close they had gotten to their point of aim for the landing. When the result came in, they were astonished. They were right on target. The accuracy of the landing impressed people all the way to the top of the NASA food chain, one of whom characterized the feat as making a hole in one with a golf ball launched in Washington, DC, toward a moving target in Australia. The pinpoint arrival was confirmed by an image snapped by the Mars Reconnaissance Orbiter, which spotted not only the lander but also the parachute resting nearby.

Soon it was time for the first look at the surrounding terrain—and what a delight that turned out to be. All around the lander were fascinating polygonal shapes etched into the permafrost. The cracks in the soil appeared to be fresh—older ones would have filled in or eroded away. This indicated ongoing changes within the soil, thawing and refreezing as the ambient temperatures swung from one extreme to another. Of course, this also highlighted the urgency of getting Phoenix's work started as quickly as possible, for these polygons were a stark reminder of temperature extremes what would kill a lander. Already, the onboard thermometers were beginning to measure temperatures that would range from –22°F to –122°F. And while Phoenix was designed to operate within this range, and colder, the specter of the advancing north polar winter, about three months away, weighed heavily on everyone involved. Phoenix would most likely not survive the long Martian polar night.

Then, just as things were getting interesting, Phoenix went silent. The link between the lander and the MRO orbiting overhead failed, plunging the mission into hours of tense data darkness. A day later, the issues had been ironed out, and the Mars Odyssey orbiter, still operating after almost a decade, was also pressed into relay service. But another snag occurred almost immediately. There was an excruciating delay while issues with the robotic arm were worked out. The low-cost approach to building and operating Phoenix seemed to be manifesting gremlins. Ultimately, the arm was unable to touch Martian soil until May 31, almost a week after touchdown.

Without further delay a sample was scooped up from the frozen soil and then the arm folded back and dumped it into the funnel leading to the TEGA analyzer. At least, that's what they thought it had done. The telemetry was less convincing—it appeared that none of the soil retrieved had actually gotten into the sample container below. Images of the entry area of the unit showed the soil sitting atop the screen that was supposed to sift the dirt before it dropped into the TEGA's oven. There the sample sat, comfy as a fat cat on a pillow. Hypotheses were formulated: the dirt was too wet, or to clumpy. It might be a clay of fine, sticky particles—not dry, loose particles. Or the screen was clogged. Or the Great Galactic Ghoul had stepped in. Nobody knew.

So controllers turned on a vibrator attached to the screen to shake some of the soil down into the TEGA. Nothing happened. They shook it some more . . . and some more. Ultimately, they shook the screen far longer than it had been designed to do, almost an hour, but still, no dirt.

There was a photocell—an electronic component that measured light in each of these small containers. If the sample had gone in, it should have blocked the photocell, at least partially. And at this point, the cell was picking up a nice, fat signal from the LED-generated light just a fraction of an inch away, across the empty container. So, no dirt.

This was one of those sweaty moments in a robotic mission: planners had to decide how to proceed with a compromised result on the lander and a collection of unattractive options. They could keep trying to get the sample down into the TEGA, but it was almost certainly dried-out and less interesting by now. Another option was to simply button up the oven and bake whatever might be in there, but if it was empty, this would waste an oven—they were strictly a one-shot deal. Or an attempt could be made to dump another sample on top of the one that was not entering the oven, but that might not work either. It was a vexing problem.

A parallel issue was an ever-expanding discussion about which area near the lander to sample next, assuming the sampling procedures could be worked out. Some scientists wanted to dig from the same trench, some wanted to try another area. And on top of it all was the communication delay and programming time—

things thought up on Thursday were not able to be executed till Friday or later.

It was like swimming through a molasses of variables. And, at ten days into a ninety-day mission, time was of critical and finite quantity. Since this lander was near the frigid pole, the usual number of mission extensions was unlikely. Engineers and specialists swarmed the robotic twin of Phoenix at the control center at the University of Arizona, seeking answers to their questions. One plan that met approval was the idea of using a "sprinkle" technique instead of a "dump" technique. Yes, it gets that precise. Rather than simply dropping the soil sample on top of the entry to the ovens, they would hold the arm's dirt scoop over the funnel and run the small rasp motor attached to it. The vibration from this should cause the fine, silt-like particles that appeared in the pictures to rattle down a small depression in the shovel itself. The result *should* be that rather than simply dumping the entire scoop of soil onto the grating, only the finer particles would be sent in and assured to enter (and fill) the oven. Or so they hoped.

The idea was tested first on Earth, then on Mars. A small amount of soil was sprinkled onto the deck of the lander . . . and success! It appeared that the sprinkle technique would separate out the finer grains from the gunky soil. The maneuver was repeated to get a sample into another one of the TEGA ovens, and things were back on track. The shake-then-bake was under way. The oven would heat the soil for days, one step at a time, ending up at about 1800°F. Then the TEGA itself would sniff for compounds in the resultant gasses.

One problem down. Meanwhile, the Phoenix team was *still* seeking consensus on the next place to sample. There were now a variety of trenches dug by the arm to pick from: Dodo, Goldilocks, and Wonderland were just three of the names used. It's interesting to compare this naming scheme to Pathfinder, when a rock could be named after Scooby-Doo® or any other critter that struck a collective funny bone. No more. One morning, someone at the NASA legal office woke up and realized that they could potentially be sued for using names of copyrighted cartoon (and other) characters. So the edict went out: use names and titles in the public domain only. Which means they have to be out of copyright. Which means waiting for seventy years after the author or creator of the work has died. Which means . . . oh, just pick a very old book. *Alice in Wonderland* will certainly do.

As the TEGA did its bake, the Phoenix team was looking at these trenches in the ground. One of the most attractive targets were bits of "white material"—which is the term the scientists used to make it clear that these were not yet identified as water ice, the holy grail—seen in some trenches. These were monitored with care to see if they would shrink over time, which would indicate melting (or more properly sublimating—going right from ice to vapor, skipping the liquid water phase). Meanwhile, the TEGA results came in: carbon dioxide, check. Water vapor, check. But no ice. This was not a deal breaker, as the sample that was finally baked had sat for some time in the scoop before being ingested into the TEGA oven, and any water ice would have vanished in that time. So the jury was still out on water ice.

Then day 23 (or more properly, sol 23, as this indicates the longer Martian day) comes along, and with it, a by-now-familiar problem in the Mars spacecraft family. Phoenix drops out of contact—the lander goes into safe mode. Data gathered in the last twenty-four hours is gone, and the lander is waiting for binary CPR. Corrections are swiftly made to the software and uplinked to Phoenix at the next opportunity; things seem fine for now. Back to the "white stuff."

After a few more days of observation, the white chunks have vanished. It was not Phoenix's doing; the lander was napping for part of the time. Bottom line: it must be ice. The announcement is made at the next press briefing.

About this time, the first results of the MECA experiment come out: Martian soil could, in fact, support some earthly crops, at least in theory. It is very alkaline, and might support asparagus but not strawberries. But that's enough.

A new problem arises: there is a short circuit in the TEGA that could affect its operation. This may or may not have been related to the excess vibrating of the sample screen. Either way, the decision is made to stop examining the most promising soil samples and go directly for a water-ice sample, just in case the entire thing stops working. This decision comes right from the top at NASA HQ.

But where to gather the icy sample? The debate on sampling-site selection seems endless. Each trench, each dig area near the lander has its own issues. Some have to do with hardness, as the sample needs to be scraped from hard ice with the drill-like rasp on the sampler arm. Others have to do with positioning: if the ice is embedded in the wall of a trench, it's harder to get at, and so forth, and so on. Regardless, a decision must be made. And there is another consideration: some of the ice nearby seems to be deposited by wind and weather over time (it is, after all, near the surface in a polygon that is formed by thermal changes over time); other ice would appear to have been sheltered and locally formed (as with the white bits found beneath the surface). Mission planners must decide which they prefer, and quickly. It is a battle between sure bets and long shots, between safe procedures and good science. And the mission is very nearly half over . . . sol 45 is now fast approaching.

It is about this time that the tests they have been conducting with the sample arm—by scraping at the presumed water ice that NASA so dearly wants—indicate that this is not an effective way to get ice samples. They will need to be harvested with the rasp on the end of the sampler arm. This will require up to five days/sols to complete. The midpoint of the mission is upon them, and the attention of many is refocused on other science results like atmospherics, meteorology, and imaging. But time is a-wasting.

Then, three instruments—the robotic arm, its camera, and the TEGA all get uppity on the same day and go into safe mode. They seem to want a day off. In the end, it turns out that the arm ran into a rock while it was carrying out an experiment. Now, when we say "ran into," remember that this thing moves v-e-r-y slowly . . . so it's more like it gently nudged a rock. But when you have a half a billion dollars worth of spacecraft working up on Mars, you tend to be careful. So the computer onboard shut it down.

The procedures are reviewed, new commands are sent skyward, and Phoenix is ready for action once again. On sol 52, they are finally ready to get some ice. The arm is pushed up against the suspect white material . . . the rasp is initiated, and starts spinning . . . and it works. The icy soil, harder than concrete, had met its match. But this is just a test. A few more sols pass before the real thing takes place.

Finally, at the two-thirds point of the mission, with the clock ticking loudly, some Slurpee®-quality ice slush is transported into the TEGA oven. Things look good. But then the "Oven Full" sensor, that pesky photocell, says, "Nope—not so." There does not appear to be enough material to heat for a reliable result.

It turns out that the ice samples were sticking to the scoop. This polar mission was turning out to be much harder than doing similar work elsewhere on Mars, and that's already hard enough. It took a third try, on sol 63, before success. The sample was delivered . . . the TEGA baked it . . . and the result: water. H_2O. *Agua*. They found part of what they came for, and all (apparently) that NASA HQ sought. There is water on the surface of Mars—and that's a fact.

Soon the other instruments had been fed and given a chance to do their analyses. It turned out that there was perchlorate in the soil after all, just as the "no-life" hypothesizers had maintained while interpreting the confusing life-science findings of Viking. Perchlorate, that nasty, reactive, chlorine-riddled, water-polluting, and rocket-fuel-making chemical was present in the icy soil. This was, like the water story, big news.

Finally, near the end of the mission, calcium carbonate (the same stuff that TUMS® is made of) was discovered in the soil surrounding Phoenix. And it's a compound formed in the presence of water . . .

Phoenix lasted 155 sols, a bit less than double the ninety-day primary mission. At that point, Martian winter was settling in, intense cold was seeping into the circuitry, and there was simply not enough sunlight reaching the solar panels to power the lander. It went into safe mode, shutting down systems and entering a state of hibernation. Not long afterward, one of JPL's orbiters spotted Phoenix in midwinter, completely encased in dry ice. It was later noted that the solar panels appeared to have cracked from the cold and the weight of the icy buildup.

After the long Martian winter, about two Earth years later, JPL attempted to contact Phoenix, just in case it had survived the long, dark night. There was no response. But this would have been merely a bonus; the plucky machine had succeeded beyond its designers' expectations already, met its mission objectives, and made a number of major discoveries.

Two years earlier, in its final message to Earth, just before it began shutdown procedures, Phoenix flashed a last bit of binary code for anyone who could decode it to hear. It was a defiant last statement as it headed into the cold polar night . . .

"01010100 01110010 01101001 01110101 01101101 01110000 01101000"

That's binary code for "TRIUMPH." A bittersweet good-bye from a small, inexpensive, and very successful machine.

CHAPTER 26

PETER SMITH

POLAR EXPLORER

Tucson is not the kind of place where you expect a polar explorer to live. And from his compound out in the Arizona desert, surrounded by saguaro cactus and scrub, Peter Smith's view is the, um, polar opposite of the icy environs he has spent time exploring (Smith pulled a stint in Antarctica testing protocols for Phoenix and spent time in the Arctic as well). But perhaps the soil is sufficiently ruddy to bring some familiarity to the scene. Or perhaps it's because he didn't start his planetary-science life with the arid deserts of Mars but with an even more frigid place.

"I was working with a professor named [Martin] Tomasko, an A-level scientist at the University of Arizona, and we were studying the atmosphere of Titan, Saturn's mysterious moon. It has a thick, cloudy atmosphere, so the surface had never been seen. Was it earthlike? Did it have weather? Did it have methane rain? There were lots of questions and few answers.

"We were making computer models of Titan's atmosphere to match the data collected by Pioneer and Voyager flybys, but it was very difficult to bring those models into a conclusion. Deep in the atmosphere, we had no idea what was happening; we were just making guesses. So we proposed to build and operate a camera for the new Cassini mission, which included a lander called the Huygens probe built by the Europeans and designed to descend to the surface of Titan on a parachute.

"Our proposal activities began in 1989, and Huygens actually landed on Titan in January 2005, so it was a sixteen-year commitment to solve this mystery about what was going on in the lower atmosphere and surface of Titan. Dr. Tomasko was quite enthusiastic about doing this and inspired my enthusiasm. My background had been in optical engineering, so I became project manager once our proposal had been accepted, and spent a lot of time at Lockheed Martin in Denver where the camera, now called DISR [the Descent Imager/Spectral Radiometer], was being designed and built. JPL managed our contract with NASA, and representatives from ESA [the European Space Agency] provided the interface to the Huygens probe.

"Scientist to manager was a difficult transition, because contract negotiation, program-management techniques, and quality-assurance methods aren't taught in school. Frankly, sitting in a room with Lockheed Martin's contract lawyers, Marty and I felt like sheep about to be shorn. Sometimes we would make *baaaa* noises when contract topics were particularly arcane yet potentially vitally important."[1]

But despite Smith's modesty, the mission was a success, and new opportunities were on the horizon: "After a few years working on the DISR [Titan] project, there was a NASA announcement of an opportunity to build a camera for Pathfinder, to be the first Mars lander since Viking twenty years before. I had never written a paper about Mars, I hadn't read a book about Mars, I knew essentially *nothing* about Mars, and yet I knew how to build cameras. We had numerous parts that we were assembling into the descent imager, including lenses, detectors, and electronic control circuits. A subset of these parts became the basis of a Mars camera proposal. Lo and behold, it won the competition over cameras designed by the top Mars scientists in the country. . . . This dark-horse victory allowed me to leave the long-term Titan project and become a principal investigator.

"After building a prototype at Lockheed Martin, we brought the project to Tucson at the University of Arizona. Now, understand that we had never actually built any flight hardware before, traditionally considered an engineering challenge outside the ability of a university, but we bucked the trend and did it.

"After a thrilling landing on the Martian surface on July 4, 1997, we had a huge success tracking the Sojourner rover as it drove around the spacecraft, and we got tremendous response from the public. It was really fun, and of course I got pushed out from behind the scenes and on stage at the press conferences. Again, this requires talents that you don't learn in school.

"So I had ideas of how a press conference should be. You know, NASA didn't do the best job of promoting these missions, they are talking heads behind a long desk and they can be very dry, they rarely show any emotion. So I wrote up this blurb, assuming it would be successful, and I was the first speaker at the press conference."

Seated next to him at the press offering was Daniel Goldin, the then NASA administrator. Goldin may have had his own ideas about how a press conference for a major planetary mission should be run, but if so, he didn't share them with Smith: "With Goldin seated right next to me, I started describing the mission as a trip to Mars like a real person would experience it. My camera had a personality. But it was not able to get the best seats on this mission and went economy class. With the parachute crammed down the back of its neck, and a solar panel pressing against its face, the trip was very uncomfortable crushed in between all these components. It took months to get there, then all of a sudden bouncing along on the ground on airbags, *bam-bam-bam*, finally stopping as the airbags deflate. The solar panels unfold and now the camera named IMP is finally able to raise its head; this is what he saw. And with that introduction, I showed the first panorama taken by the camera.

"So I'm giving that kind of a talk, from the point of view of the camera . . . as I'm halfway through, I look over at Dan Goldin and he's glaring at me like 'You are out of business . . . ,' so now

I'm thinking, oh my god, maybe I shouldn't be doing it this way . . . but the press was enjoying it and we were having fun. Then I showed the first image from the surface of Mars in twenty years and they were really excited, we got a big standing ovation! It was fun and I was really enjoying my new role as camera guy bringing pictures of Mars to Earth.

"The next time I saw Goldin was at a planetary-science meeting. . . . [H]e recognized me in the audience, so as he left the stage he came right across the room to me and shook my hand. I guess he remembered me for my public-relations ideas. Mission success is what it really takes."

It was at this moment that Smith realized what space exploration was about for him: "It was exciting for people everywhere; I felt that this was one of the positive things we do in America and give away freely to the world, and so I was very excited about contributing in a small way. Our weather satellites, monitoring of the environment, exploring the planets, all of this is what NASA does for the world and what we as American taxpayers have supported for fifty years. We develop new technology, encourage our young people—it's such a positive message, and that's why I love working for NASA. People need to understand, we don't ship $250 million into space, it's all spent right here in our universities, NASA centers, and aerospace companies. The data advances the science and our knowledge of our solar system. That's where the money goes."

It is said that no good deed goes unpunished, and NASA's next foray to Mars would seem to have proved this maxim: "The next mission was the Mars Polar Lander [MPL], and I had built another camera system [for it]; it launched in early 1999 with a landing scheduled for December 3, 1999. On the day of landing, it separated from the cruise stage, went into the Martian atmosphere, and was never heard from again. What could we do?"

Between the loss of the Mars Polar Lander and the Mars Climate Orbiter (MCO), NASA and JPL were due for a period of intense internal review.

"I had to fire all the people that worked for me; the mission was over. NASA canceled the next mission, the 2001 Surveyor, on which I was an instrument scientist. My team shrank from thirty-five employees down to one employee. So I limped along for a couple of years until I could propose Phoenix, and Phoenix was the rebirth of the 2001 Surveyor Lander that was mothballed because of these crashes. I had experience with the lander and was a natural PI [principal investigator] for the mission, although frankly, it wasn't even my idea."

It was not a pleasant time for anyone involved, but they would rise from the ashes of defeat, and brilliantly. The idea came from Dr. Chris McKay, a Mars researcher from NASA's Ames Research Center: "What can be done with the hardware which remains at JPL from the [canceled] Mars Surveyor Lander mission?" he asked.

"We had the Surveyor spacecraft and the instruments that were delivered for its mission. We had a launch date. But we had no science goals, so I had a few weeks to provide the vision that would make the mission attractive to the science review board. This was exciting stuff—what if you could do any mission that you wanted to? I'm just a lucky guy, able to propose a mission to the surface of Mars. My mind went into high gear.

"So many possibilities; I thought about the chances for finding life, about where I would look, under what rock, how would we get there, and what would we do. What instruments would we need? That exact week, one of the professors in my department published a paper about finding ice under the northern plains of Mars. Even though you see it in camera pictures as a dry, dusty plain, he could probe under the surface using gamma rays and neutrons, and could see down about a meter, and his team found that there was a solid ice layer right under that surface. He did this with the Mars Odyssey orbiter, using the gamma-ray spectrometer."

Smith recognized the opportunity immediately. He had just been handed the keys to an exciting mission—a newly conceived look at the icy polar region of Mars, a likely candidate for past and possibly present life. "So, they had just nailed it, found all this ice right near the surface! If we could go there and understand the history of that ice, and the minerals and the chemicals that are in association with it, that would be an incredible mission. Could that be a place like here on the Earth, where you find these Mars analogs, like in Antarctica and in the permafrost regions of the Earth? There you can find evidence of past life going back millions of years. The permafrost is the deep freezer of the earth. That's where living things are preserved. They aren't preserved in the jungles of the equator where decomposition acts very quickly.

"I once heard a talk from an Austrian scientist who sounded just like Arnold Schwarzenegger . . . [Smith lapses into an Austrian accent] 'Ve gather a cubic centimeter of soil and ve put in into our DNA analyzer, and ve can re-create zee tree of life from this piece of soil from zee Siberian perr-mafrost.' I was amazed—to think that you can re-create the tree of life for the whole Earth from one chunk of Siberian soil, just because it is all preserved there! The winds carry spores, pollens, and microbes of all sorts, and it's frozen into the soil along with the evidence of Siberian creatures. I thought maybe this happened on Mars, and so this might be a good place to look at the history of the Martian permafrost. So we wrote our mission goals around the permafrost of the northern plains of Mars.

"We chose our landing site where the strongest ice signature was seen. We wanted to understand the chemistry and mineralogy of the soil that was in contact with the ice, and to see if that ice had ever melted, because water is a very powerful agent of change chemically and mineralogically. The transition from unmodified volcanic soils that have never seen water to the altered minerals—clays, sulfates, and carbonates—produced in wet environments has been well studied by geologists. So we were going to look for altered minerals, and the instruments were picked just for that purpose. What about any organics associated with the ice? To answer that, we had to get a chunk of ice into our instruments and use it to make that determination."

In May 2008, Phoenix had made its way to Mars and was preparing to land. But after the twin failures of MCO and MPL, tensions were high: "A lot of the managers at JPL were fixated on what would happen to future missions if the Phoenix mission, which of course was a low-cost 'Scout' mission, were to fail. They felt that it could bring into doubt JPL's ability to land safely on Mars. The unknowns and what-ifs could lead one to imagining terrible outcomes; they worked themselves into such a lather. As a result, we spent a considerable amount of time preparing a press conference for failure. It was so disturbing. We actually rehearsed a press conference revealing how we had crashed on the surface because the parachute had failed.

"So, as we got to the actual day of landing, the press office had written out press releases so that they would be fully prepared after we found out what the failure mode was! We had releases for every likely failure mode, perhaps half a dozen of these things.

"As we are coming down and going through the different phases of the landing, and the failures hadn't happened, the press agent is tearing up and throwing those failure briefings into the air like confetti! I was happier and happier. So finally we got safely to the surface, and, despite predictions, two hours later the first pictures came down. Contrary to some predictions that these first pictures may not show much . . . they were just spectacular. That to me was one of the great highlights."

Once on the ground, a communication error delayed the opening of the robotic arm. But it was worth the wait: "About a week later, we finally got the robotic arm working. We wanted to make sure that the landing feet were firmly anchored on the surface, that we weren't tilted up on a rock or something, so we looked under the lander and we saw that the thrusters had blown away the soil and exposed the ice that we had come to find! Unfortunately, the robotic arm couldn't reach under the spacecraft, but it could take pictures with its camera. So here we are thinking, oh my gosh, what if the ice is only under the lander and not beside it where we need it to be? Nonetheless, it was exciting, because we were pretty sure that we were going to find ice at our site. We already knew there was ice in the northern plains, but the spatial resolution was very poor in those maps, so we didn't know if there would be ice exactly where we landed."

But, of course, there was. The results were nothing short of amazing.

"The final thing that was totally astounding and unexpected was the chemistry experiment identifying a chemical called perchlorate. Now, I am personally not a chemist, and I didn't know much about perchlorate, so my first thought was that it was a bleach, chlorine bleach, and I thought, 'My god, the soils of Mars are filled with bleach, we're never going to find life here!' I mean, that's how you clean the microbes out of your drinking water, right? But it turns out that the chlorine in the perchlorate form is very stable, very soluble in water, and has four interesting properties. One is that it lowers the freezing point of water, so if you concentrate perchlorate, you get a very low freezing point in brines, and life could survive in those brines. Second, there are microbes on Earth that live on perchlorate. They use it as a food source, an energy source. Third, perchlorate is used in solid rocket fuel, so it could be a resource for astronauts. The fourth property is that in high concentrations it is toxic to humans—future astronauts will need to be prepared to protect themselves. It was all very exciting."

The best was yet to come. Phoenix went on to not only be the first mission to successfully land on the polar regions of Mars, but one of the more successful static landers overall, despite its short life.

When asked to reflect on his accomplishments, Smith leans back in his chair. The answer is not what one might expect from one of America's top planetary explorers: "It's simple. I'm a lucky man: lucky to be able to work with some of the best scientists and engineers in the world. Together we accomplish great deeds."

It's more than luck, as anyone involved who has observed his hard work knows. But don't tell Peter Smith . . . he's busy designing instruments for NASA's new asteroid-rendezvous mission. And he is likely hoping for luck on that one too.

CHAPTER 27

MARS SCIENCE LABORATORY

BIGGER IS BETTER

If Pathfinder's Sojourner rover was a remote-controlled toy and the Mars Exploration Rovers were ATVs, then NASA's new Mars Science Laboratory (MSL) rover is an SUV. The mammoth mobile science platform is the newest Mars effort, launched on November 26, 2011. And it's a doozy.

Not since the Viking landers has anything so large and complex been sent to the surface of Mars. Nuclear-powered and better-equipped than any of its predecessors, MSL is set to redefine Mars-surface exploration in a big way. Just the numbers alone are impressive:

- Where the MER rover weighed about four hundred pounds, MSL weighs two thousand.
- MER's instrument package weighed less than fifteen pounds, MSL's weighs 276.
- The Spirit and Opportunity rovers were just over five feet long, MSL is over ten.
- MSL can surmount obstacles almost a yard high, well over double that of MER.
- MSL has a minimum expected travel range of over twelve miles during its two-year mission, far longer than the initially expected range of MER.
- MSL's heat shield is the largest ever flown, larger than the Apollo heat shield of the 1960s.

- MSL is powered by a plutonium heating element, one of NASA's last,[1] generating five times the power of MER's solar panels, and it will not suffer the power reduction that Sojourner and MER suffered during periods of reduced illumination on their solar panels, with a life of over fourteen years.

MSL will not bounce to a generally defined landing zone within a cocoon of airbags as previous rovers have. The craft needs to make a pinpoint landing. This rover, the size of a Mini Cooper® automobile, is also far too large and heavy for the beach-ball method, and herein lies one of the biggest headaches of the mission. It will come into the Martian atmosphere much as other landers have—at high speed, surfing on its heat shield. But once it nears the surface and ejects its protective coverings, and after being slowed by the usual parachute (albeit a far larger one), eight rocket motors will brake the spacecraft to a hover, and the rover will be winched down from an array of cables from which it will dangle. Once the rover itself reaches the surface, the cables (also known as "bridles") will be cut with pyrotechnic charges, and the descent stage will drift away to crash nearby.

The system is known as Skycrane, and that's how it's *supposed* to work.

It is easily the most complex system ever devised for a Mars landing, and with that comes exponential amounts of complication and many grueling tests. Even the atmospheric-entry phase is more complex, and the targeting for this landing is far more demanding and tightly aimed than its predecessors. As it enters the upper atmosphere, the brain of MSL will compute the amount of deviation between its proposed course and its actual location in the air. By throwing off small amounts of ballast during this high-speed entry phase, the spacecraft will be able to steer through the Martian skies, resulting—hopefully—in a relatively pinpoint landing. Again, it's a white-knuckler. But that, as has been said, is the Mars-exploration business.

Testing has been the key to success with the Mars landers, and MSL is no exception. In fact, the testing regimen for this mission may be the most involved yet. One example is the landing radar software. A mock-up of the computer onboard MSL was placed in an F-18 fighter jet and lofted above the Mojave Desert in Southern California. The aircraft climbed to forty thousand feet and began a series of dives at angles ranging from forty to ninety degrees. Each time, the jet pulled out of the dive at about five thousand feet. During the dives, the computer checked against the radar, allowing for a simulation of the mid-altitude descent of MSL through the Martian atmosphere. This allowed the designers and programmers to fine-tune the package to provide the most accurate landing they can.

The rover's given name is Curiosity, selected from thousands of submissions from schoolchildren nationwide, and it carries the largest and most complex scientific investigative package ever landed on another world—including the Apollo lunar missions.

Investigative goals include examining the geology and climate of the region it surveys as well as continuing the search for life-compatible environs and organic compounds, if any exist. And of course, the rover will search out possible evidence of past—and with luck, present—water.

The rover has a mast with cameras at the top, as all Mars rovers have had. The cameras are capable of high-resolution stereo (3-D) stills, as well as HD video, a first. And, in another first for rover cameras, the Mastcam has a ten-power zoom lens. The cameras for this mission are being built by a private company, Malin Space Science Systems, which has traditionally built orbiter cameras. Along with the usual engineers and scientists on the team, movie director James Cameron was brought into the mix for his creative and technical abilities.

Another camera, the creatively named Mars Hand Lens Imager (MAHLI), will be mounted on the robotic arm (also a staple of all rovers) to take microscopic images of the rocks being investigated.

It has both white- and UV-light sources to see differing aspects of the rocks.

A descent imager, the Mars Descent Imager (MARDI), capable of saving four thousand rapid-fire images, will document the trip down to the surface. This will allow for mapping the area in which the rover lands in great detail as it approaches, not unlike the Ranger moon missions of yore (but hopefully with a softer landing!).

The amazing ChemCam is a suite of instruments that includes the first laser-induced breakdown spectroscopy device, which can target a rock from about twenty feet and vaporize a sample, collecting a spectral sample as it does so. The laser used is a 10-megawatt beast that is perfect for vaporizing bits of rock and sand, but would not be friendly to any Martians nearby (we should be so lucky!). This instrument is a combined effort of the United States and France.

Like Sojourner and the MER mission before it, Curiosity will carry an alpha proton X-ray spectrometer (APXS), which hails from a consortium of Canadian and US universities. This instrument can identify other minerals in the targeted sample.

A unique experiment called Sample Analysis at Mars (SAM) is an echo of the Viking life-science lab and can investigate samples both externally and in internal containers à la Viking and Phoenix. It uses a mass spectrometer, a gas chromatograph, and a tunable laser spectrometer to seek organic compounds. The latter instrument can, unlike Viking, differentiate between organic and inorganic substances. If a lucky moment occurs, this may settle quite a few arguments about the nature of Martian soil and its constituents, while inevitably raising new questions. This is another US/France collaboration.

The Russian Federation has supplied the Dynamic Albedo of Neutrons device (DAN), which will seek to indentify hydrogen, ice, and water at the surface. Another device, the Radiation Assessment Detector (RAD) will investigate near-surface radiation levels in an effort to determine necessary protection for future human explorers of Mars. To gain a better understanding of the weather such explorers might face is the Rover Environmental Monitoring Station (REMS), a meteorological suite designed to measure Martian weather.

Another suite of devices will measure the environment the lander passes through during its descent and is called the MSL Entry Descent and Landing Instrumentation (MEDLI). Finally, the obligatory navigation and hazard-avoidance cameras will adorn the corners of the rover, identifying formations to be avoided and building 3-D maps for autonomous navigation across the Martian surface.

Coordinating and commanding the instruments and the rover itself is the onboard computer, once again the Power PC© chips used on so many preceding craft. Two of the RAD750 chips will power the computer, one assisting the other. The computer is a vast leap over those used in previous probes in terms of memory capabilities, and it is over ten times faster. The main computer is assisted by an inertial navigation unit. This is not unlike that used in the Apollo program, and is again similar to the accelerometers and related measurement system used in today's iPhones®.[2] It allows the rover to measure its current location by comparing speed and direction changes since its last known location.

After years of debate and consideration, a region known as Gale Crater was chosen as the landing site from about fifty other possible sites. The crater is almost one hundred miles wide, with a central mountainous peak rising about three miles from its center. The twelve-mile-wide landing zone places Curiosity near this peak. The rocks in the region are of a different appearance than those investigated by other rovers. Much of what will be found there should be rocks that have tumbled down from the crater wall as well as from the central peak. Additionally, as with other craters, there will be a bonanza of outcrops and strata visible to the rover in the walls of the crater, exposing many millions of years of geology. Add this to Curiosity's ability to use its laser spectrometer to do remote sampling of rocks, and it's easy to see that this mission should be one for the record books.

Near the base of the central peak is an area that has been identified from orbit as rich in clay and sulfates, both formed in the presence of water. As always, this is a prime target for investigation—but especially so in Curiosity's case, as it will have the ability to sniff out organic compounds. When Curiosity does find a desirable sample, controllers have many choices of how to proceed. The sampling arm, over five feet long, has a full complement of collection options available. In an evolution over the MER sampling techniques, this rover has a small rock drill, which will allow it to grind powder from rocks. The drill has an interchangeable bit on the end; if a bit becomes dull or stuck in a rock, Curiosity can simply eject it and pick up another drill bit from a collection it will carry onboard. It also has a rotary brush for cleaning rock surfaces, similar to the RAT on MER.

Also positioned at the end of the sample arm are collection sieves and containers. The samples, whether they be sand, drill-generated powders, or other materials, can be sifted into a scoop on the end of the arm and then delivered to the onboard lab for analysis. And the rock drill itself actually has the capability to transport the powders it generates along the drill, up out of the hole, and into the sample container.

MSL has not come cheap. The overall cost of the mission is expected to be about $2.5 billion over the life of the mission. This includes massive cost overages across the last few years, and there will probably be more by the time the extended missions are complete. But there is more to this story than overall cost.

For one thing, all space missions, especially the manned ones, tend to be underestimated at the outset. This is not a lack of foresight or planning on the part of the folks who price-out the mission. Rather, it is a combination on unforeseeable price changes and, more often, the need to quote low in order to get a mission approved. In other words, one must assume the best possible outcomes in terms of cost to pry the funds from an increasingly frugal federal government.

It should be noted here that JPL's endeavors have almost always returned many, many times what has been promised. Orbiters designed to last a year or two routinely continue operations for five or even ten years. Landers and rovers expected to operate for a few months end up running for years and years—and in MER's case, often covering ten times or more than the expected territory. And this is not due to deliberately lowered expectations on the part of mission planners. The mission objectives are specified by NASA headquarters, and it is expected that each effort will meet these minimum requirements. After that, the rest is frosting on the cake. And JPL's missions traditionally produce a lot of frosting; it's like finding thirty pounds of buttercream on a cupcake.

Curiosity is scheduled to land on Mars in August 2012. Standby.

CHAPTER 28

DR. JOY CRISP, MARS SCIENCE LABORATORY

DIG THIS

Married to another JPL scientist, Joy Crisp can be found on off hours at her Princeton home quietly immersed in a science fiction book, often a David Brin title. She has been at work on the Mars Science Laboratory for years, acting as the deputy project scientist, but her path to Mars was not a simple one.

"I was a volcanologist, so I studied volcanoes on the Earth. I was doing a postdoc at UCLA, and a friend of mine said 'I think there are people at JPL that are volcanologists, and there might be a postdoc position open there.' I had no idea! I thought JPL was just a place where they studied space. I talked to them and sure enough they were using thermal infrared sensors to look at Hawaii. I started to do research at JPL with a group of people, and then Pathfinder came along and they needed someone who could work on the instruments like the APXS [alpha proton x-ray spectrometer], which measured the chemistry of rocks and minerals, and they said they needed someone that knew about geochemistry. I was one of the few people that had that expertise, so I got involved and it was interesting. I moved to the Mars Exploration Rover project, and now I'm on Mars Science Laboratory. So I transitioned from studying volcanoes on Earth to volcanoes on Mars, and I ended up doing all kinds of projects."[1]

Of those projects, the Mars Science Laboratory is easily the largest to date.

"This is a very big project, so there's a lot of things to do. We must make sure the science team can carry out their investigations, keep an eye out for the things the engineers are doing that could affect science, and advise a project manager when he has to make decisions.

"[MSL] really is a stepping-stone beyond missions like Pathfinder and [MER], which were very geology focused and didn't really have any capability for looking at organic compounds and the building blocks for life. [The] Mars Science Laboratory is better equipped.

"Pathfinder was a technology-demonstration mission. It was a short-lived mission and we confirmed a lot of things we knew about Mars. We did measure some slightly higher silicon composition, so there were some ground truths right at the site where we landed. Spirit and Opportunity were a huge step in understanding because they were more capable, and because they lived so long. Opportunity is still going, and because of that, we've learned a tremendous amount at two very different sites. One thing that those rovers have done is to show us that there is definitely a diversity of geology on the planet and that we can redirect [the rovers] and find evidence of past water. With Opportunity, we found some rock layers where water was even flowing on the surface, depositing the grains, and also secondary water, ground water, was circulating through them, and cementing them, and making those little hematite spheres in them. So there were lots of clues.

"What we will really be able to do much better with Curiosity is to identify the minerals. We were struggling a little bit with Spirit and Opportunity; we could identify iron-bearing minerals with the Mössbauer spectrometer, but with other minerals, we had to guess somewhat. [We took clues] from a thermal infrared spectrometer as to what mineral combinations might be there. So when we get there with our x-ray diffraction spectrometer on Curiosity, we'll have a much better way to say what minerals are present in the soil and in the rocks that we look at.

"We're also bringing the instrument called SAM, Sample Analysis at Mars; and that one will be able to drill into rocks and find out [if any] organic compounds are present. We haven't tried to do that since Viking days, and when Viking tried to do that, it couldn't find any organics in the soil. We're going to have a more sensitive instrument. We'll be able to heat [the soil] up much higher and be able to look for organics at even a lower level and look at drilled rocks. It's still going to be pretty hard, and it's a remote possibility that we're going to find organic compounds on Mars, but we'll certainly have a better chance of doing it with this rover."

The MSL rover is not designed to search for life, though. It will search for the basic elements that can support life: "We're not trying to do what we did with Viking, which was to look for life. [With Viking] after we got the experimental results, we scratched our heads and realized that we could think of a way for an inorganic substance to create those kinds of results. That wasn't the best test, and we realized how hard it would be to devise an experiment to look for life. We don't have an instrument that the science community can [agree on] to go look for life. So we're kind of taking a baby step in that direction, going slower than Viking tried, saying 'let's try to find the organic compounds and measure those again in a better fashion.'"

So . . . is there life on Mars? The answer is unclear, but Crisp can hazard a guess: "We believe it's more likely that there was life in the past than life today because of the harsh environment today, but we're still going to go and . . . drill into a rock five centimeters [(two inches) to see if we] find organic compounds preserved in the rocks. We're trying to use techniques that we use on the Earth to look at the rocks and say 'which one of these are most likely to preserve evidence of organic materials?'"

To collect these samples, MSL will use traditional, tried-and-true techniques, such as a sampler arm with a scoop and rock brush. But there is a new wrinkle in the mix: the rock drill. And getting powdered rock from the drill to the onboard lab in the rover will be yet another challenge: "This is a huge new challenge that we have not tackled before. We did a little bit of this with Phoenix, where they had us scoop and deliver material into an instrument with the wind blowing. We learned a lot of lessons from [that mission], but we're trying something even harder with the rock drill."

It's natural to assume that it must be frustrating for geologists like Crisp to work from so far away. To this, she responds: "Well, I'm a geologist, so I like to go out with a rock hammer and hit rocks and look at them. I want to know things like how did this rock form, what was it like when this rock was forming or altering, and so on. MSL is just the kind of mission that excites me; it's as if I could be there, because I'm drilling in the rocks, and then I'm finding out what minerals are in it and looking at it with a close-up camera. And this time it will be in color and higher resolution! In all of our sites we have layers of rocks so we can move through and see how things changed over time in Mars, so that's going to be interesting too."

But still . . . commanding a machine millions of miles away is far tougher than doing it yourself. And the team must be trained extensively for this: "We had a science team test where we sent some people out to Arizona. It was a site that the team didn't know [the location of], and we set up a bunch of equipment that was like what we were putting on the rover. We started out by taking pictures and we put it into their planning tools. Everybody was working from their home institution around the world, and they started out with a bunch of pictures, and we told them, "here's your picture from orbit, you are here, it's day number 235; now start planning tomorrow and here's what you were thinking of doing." Many of them have never done this before. A few of them were from the Spirit and Opportunity missions, but many of them had no idea what it would be like.

"One of the lessons learned was that we had no idea how frustrating and challenging it would be to do field geology so slowly.

When you are planning the next day's work, you have to argue with your peers on what steps to take. For instance, will the rover drive this way or that way, put up its arm or not, and so forth. We wanted those kinds of lessons to sink in so that they start getting used to it. Personally, after so many years of doing it myself, I just mentally accept that this is how it works. I'm very patient."

So, given all this, would she prefer to go do it on-site?

"I wouldn't want to go to Mars myself yet because I'm just not ready to do that, it would be way too difficult right now. So I'm willing to do it this way, slowly, via computer. It's a different kind of challenge . . . can you work with your scientist friends to come up with the best plans to get the rover to do things, and then sift through that precious data to get the most out of it that you can. It's just a different kind of challenge. Like I said, I'm a patient person."

And, as we know, patience is a virtue rewarded in planetary exploration. The secrets of Mars await.

CHAPTER 29

JPL 2020

THE ONCE AND FUTURE MARS

Jet Propulsion Laboratory is seventy-five years old as of 2011. In that time, the campus has seen jet- and rocket-engine experimentation; construction of America's first satellite, Explorer 1; missions to the moon, Mars, Venus, Mercury, the sun, Jupiter, Saturn, Uranus, Neptune, as well as the asteroids Vesta and Ceres and the comets Tempel 1 and Hartley 2. And, with the passage of the Pioneer 10 and 11 and Voyager 1 and 2 spacecraft out of the solar system's boundaries, JPL is now officially in the business of interstellar exploration as well.

Of course, no one institution could do these things alone. JPL is funded by NASA and managed by the California Institute of Technology, also in Pasadena. The many missions it operates are done in cooperation with institutions all over the country and beyond. Notable among them have been Stanford, Cornell, the University of Arizona, the University of Colorado, and many others. Space is too vast, the job too huge for one agency to go it alone.

With the Opportunity rover still operational, Mars Odyssey, the Mars Reconnaissance Orbiter, and Mars Express still sending home data, and the Mars Science Laboratory on its way to the Red Planet, what lies ahead?

Currently, besides MSL, the lab is involved with parts of the James Webb Space Telescope and other Earth-orbiting observation platforms and is cooperating with the European Space Agency on other planetary missions. A lone Scout-class mission, MAVEN, is scheduled for a possible 2013 launch. It is a small and inexpensive orbiter to study the Martian atmosphere. Beyond this . . . no other funded Mars programs exist.

There have long been plans for a sample-return mission, but this is a far larger funding requirement than mere landers and rovers, and so far NASA has not allocated the dollars necessary to design and build such a spacecraft. JPL is also working with NASA and ESA on the ExoMars mission, with an orbiter planned for 2016 and a rover for 2018, but the fate of this European mission is uncertain. And the United States would be a junior partner at any rate.[1]

So a reasonable person might ask: what does an agency like JPL need to do, beyond racking up decades of brilliant successes, most of which have far outperformed their designers' wildest fantasies, in order to secure future projects and funding? Said reasonable person might be stunned to find that such performances are not enough. The public at large, and Congress and the executive branch in particular, seem to feel that these bravura performances are the *minimum* expectation. They do not ensure future funds. And major failures, such as the Mars Climate Orbiter and Mars Polar Lander debacle, could bring down the whole show. The American public seems to have a short memory for success . . . ask any Apollo astronaut other than Neil Armstrong or Buzz Aldrin.

But there are plans. Which of them will be funded and developed remains to be seen; what follows are some of the more likely candidates for future Mars exploration.

The most exciting for most observers is the Mars sample-return mission. Long a twinkle in NASA's eye, a sample return will incorporate all the experience gained from the last twenty years of Mars landers and more. This craft must descend to a pinpoint landing, discharge a smart rover with the ability to handle larger samples, be capable of acting as a stable launch platform and then launch a rocket able to depart Mars with a load of rocks and soil and navigate back to Earth, including atmospheric reentry and landing. It's a huge and daunting undertaking, and may require international partners such as Europe, Russia, and perhaps newcomers such as India or China to succeed. But in the end, it seems likely that a successful sample return will be funded primarily by NASA and run, of course, by JPL. A Martian sample studied on Earth, with all the luxuries of a fully equipped laboratory, will yield answers to long-held questions about chemical composition, the existence of organic molecules and much, much more than could ever be accomplished robotically on-site.

NASA is also still considering a series of ongoing smaller missions. These include ideas such as Mars airplanes, large instrument–toting balloons, and more landers similar in scope to Phoenix. Both the balloon and the airplane proposals are for craft that would stay aloft in the Martian atmosphere for weeks, if not months. These airborne platforms would allow for a close-in observation of the many points of interest spotted from orbit. Originally, many such plans had fallen under the now-canceled Scout program.[2] Some may be reclassified into NASA's ongoing Discovery program.

An astrobiology-laboratory rover has long been on the drawing boards. Building on experience gleaned from MSL, such a rover would represent the first true search for life on Mars since Viking. But it would be far more sophisticated than Viking or even MSL, and would likely be tightly focused on microbial life. If MSL is successful, look for this advanced rover sometime late in the decade.

More orbiters will doubtless follow MAVEN, as there are always increases in imaging and sensory capability to exploit in a new mission. Once sufficient improvement builds up, there comes a point of critical mass that drives a new Mars orbital project. Before long, Mars orbiters should match the capabilities of current Earth-orbiting spy satellites.

Finally, further exploration of the poles and deeper Martian soils is expected. The one major class of geological investigative tools that has not been included on a flight to date is a deep-soil drill. This will be another leap in mass delivered to the Martian surface, as rock and soil drills are heavy. The technologies explored in MSL should aid in the design of this unit.

But these are in the future, and the future of space exploration is a fragile thing. It is tempting to consider JPL and NASA to be forever; to be eternal institutions. But this longevity is far from assured. With financial crises rocking the globe, and the US federal government seeking ever more ways to cut spending, there are few sacred cows. Science is never safe from funding cuts. NASA is still struggling to recover from the loss of the Constellation project to return to the moon. The space agency is left with a crew capsule, Orion, but currently has no rocket to place beneath it. All plans for a replacement launch vehicle are, at press time, far from reality. And even if funded, all NASA projects are subject to cancellation at the whim of an ever-fickle Congress.

As regards the continued investigation of the solar system, one major failure on the order of a mission such as MSL could, in the opinion of some, spell the end of JPL and unmanned exploration. More measured consideration sees darker times in such an event, but an eventual return to space by JPL in some form. But it would be a tortuous path.

Of course, this is all conjecture. But if the short history of space exploration is any guide, it is a seemingly easy item to trim, if not outright cancel, from the national agenda. And this is a shame, because the exploration of space is something the United States has consistently done better than anyone else, and it is one of the few programs in which the money spent is returned, at a rate of almost 100 percent, into the American economy. Jobs and education benefit; engineering and science are enhanced. It's a classic win-win scenario, but one that is increasingly hard to sell to the American public at large.

Time will tell.

CHAPTER 30

MARS ON EARTH

As compelling as the robotic exploration of Mars is, sometimes it takes "boots on the ground" to decipher the secrets of a new world. With human flights to Mars still in the future, some intrepid researchers have taken matters into their own hands. They have organized expeditions to Mars . . . on Earth.

As our understanding of the Red Planet has expanded, it has become clear that there are places on Earth that mimic Mars in some important ways. If this discussion were taking place in 1904, we might discuss the Mojave Desert in California or the canals of Holland. But in the twenty-first century, we understand far more about Mars and there are some surprising parallels on our own planet. Devon Island in the Arctic . . . the Atacama Desert in Chile . . . the dry valleys of Antarctica. There are more, but these are some of the primary research targets of scientists seeking a so-called Mars analog on Earth, usually to study primitive forms of life, soil conditions, or research techniques.

One of the most compelling of these endeavors is the Flashline Research Station, or FMARS, set up on Devon Island, about one thousand miles south of the geographic north pole. Run by an organization called the Mars Society, it is a cooperative venture between the society, NASA, and academia. The effort to build the station was spearheaded by Dr. Pascal Lee, an accomplished research scientist with NASA at the Ames Research Center, and Dr. Robert Zubrin, a brilliant aerospace engineer formerly at

Lockheed Martin and a founder of the Mars Society, a space-exploration advocacy group. Zubrin has long been an advocate of finding cheaper and more effective ways of traveling to Mars, for both robots and human beings. The Flashline Station has proved to be an excellent exercise in what such a mission, once landed upon the Red Planet, might entail.

Zubrin was a cofounder of the Mars Society, which has funded, has built, and operates the station. It was not a simple task. Many years of intense fundraising, multiple design studies, fabrication, and the transport of the prefab materials to Northern Canada were just the beginning. Much of the prefab elements were severely damaged before final construction, and a lot of repair work had to be done on-site on the remote, frozen landscape of Devon Island. Then a crane failed, and the final assembly needed to be carried out via old-fashioned block and tackle. The construction crew had departed, and it was up to Mars Society volunteers (including Zubrin himself) and a film crew from the Discovery Channel (who set down their cameras and joined the volunteers) to complete the structure. It was chilly, exhausting work. But by 2000, the habitat was complete and ready for the first crew.

Since 2000, crews picked from academia and industry have spent rotations at Flashline Station. This is a true simulation; the six or seven crew members must don simulated pressure suits when they work outside during EVAs (Extra Vehicular Activities), including a timed depressurize-pressurize cycle upon leaving or entering the station, or habitat, as they refer to it. Communications with the outside world are typically delayed by twenty minutes to simulate the one-way travel time of radio from Mars to Earth. The station itself is a large cylindrical unit, perched upon supports that hold it just off the icy surface. Inside are basic accommodations for a small crew, including bunks, a galley, research stations, a satellite computer link, and other common areas. The only real concession to Earth-bound logistics is the single crew member armed with a shotgun, as well as nonlethal deterrents, to guard against polar bears that might take an interest in a space-suited snack.

Once outside in the cold Arctic air, crew members either perform experiments and maintenance duties near the station or climb aboard small gas-powered ATVs to traverse to specific targets, usually for meteorological, biological, or geological activities. Daily logs are kept by each member.

The entire effort is designed to mimic as closely as possible (on a limited budget) a stay on Mars of up to a month or more. Participants have ranged from NASA scientists to university grad students to journalists.

Data from these stays have provided valuable information on psychological and logistical problems, research and experimentation techniques, and more. Of particular interest have been their studies of the region itself. The station overlooks the Haughton Impact crater, a fourteen-mile-wide site where a large extraterrestrial object slammed into the Earth some thirty-nine million years ago.

Crew members continue to serve rotations at the Flashline Station, and will continue to do so for the foreseeable future. Interested? Laypeople are now being encouraged to apply . . .

There are other ways to research Mars on Earth, however. Dr. Chris McKay is behind such an effort. After earning a doctorate in astro-geophysics from the University of Colorado in 1982, he went on to become a research scientist at NASA's Ames Research Center in California. McKay has spent time at most of the Mars analog sites, but one of the more remarkable trips, especially in terms of results, has been a journey to the Atacama Desert, one of the driest, most desolate places on Earth. McKay was prompted to explore the region by the discovery of perchlorates by Mars Phoenix in 2008. The results of the probe's experiments reopened the lingering question of the confusing life-science results from Viking, some thirty-two years earlier. And since a trip to Mars was

Figure 29.1. ARCTIC TEST TRACK: Using the Arctic and Antarctic for spacecraft testing has become a tradition for NASA. Here a remotely operated rover called 10-K ("Ten-K") is driven across the floor of the Haughton Crater on Devon Island in the Arctic. The controllers are in Northern California. *Courtesy of NASA.*

out of the question for the time being, the Atacama was the next best thing.

The desert lies at an altitude of three thousand feet and is blocked from rainfall by two bordering mountain ranges. The soil is virtually sterile and is fifty times more arid than the Mojave Desert. Misty rain drizzles onto the region an average of once every ten years. Items implanted in the soil—whether plants or microbial—die quickly. Other than atmosphere and temperature conditions, it is a near twin of Mars, and has apparently been since its formation over fifteen million years ago. It is officially the "deadest place on Earth" in its dry core region.

McKay had been to the Atacama a number of times. He took some soil from the region and repeated, as closely as possible, the Viking life-science experiments—and the controversy over Viking's results was reignited almost overnight.

Earlier research efforts had shown that the few organic substances present in the soil were so dispersed and were released at such high temperatures that, had Viking landed in the Atacama instead of Mars, the results of the experiments would have been the same: no life would have appeared, despite the fact that it existed.

In 2003, McKay discussed the results of another repeat of the Viking life-science experiments. He tested the microbial nutrient broth in the coastal region of the Atacama, where there is a bit more microbial life, and the nutrients were consumed. A variant of the broth from the Viking experiment was prepared that was not designed to support life, but was still consumable, and it was not metabolized by the microbes. But in the drier, deader inner core, both the microbe-friendly and non-microbe-friendly broths were used up equally. It appeared that the majority of the Viking scientists had been right—something else, possibly a strong oxidant like perchlorate, was reacting with the liquids.

But in a similar experiment performed after Phoenix confirmed perchlorate in Martian soil, which was still in doubt during the Viking era, organics were observed using similar instrumentation and interpreted with the foreknowledge of perchlorate in the soil. "Contrary to thirty years of perceived wisdom, Viking did detect organic materials on Mars," McKay said upon reviewing the results.[1] "It was only by having it pushed on us by Phoenix where we had no alternative but to conclude that there was perchlorate in the soil. . . . Once you realize it's there, then everything makes sense."

This conclusion is based, as mentioned, on the evidence of perchlorate in the soil. With perchlorate present, small amounts of organics could have been present on the Viking samples, but would have been destroyed in the heating process.

To strengthen the case, and this was the clincher, McKay and co-researcher Dr. Rafael Navarro-González reexamined the chemical results of the Viking experiments. Elements released from the pyrolitic experiment (the gas-chromatograph examination of gasses released by the baking of a soil sample) had shown traces of chemicals that were dismissed by the researchers of the time as Earth-based contaminants, that is, things carried up from Earth aboard the lander, despite the extensive efforts at sterilization. But when McKay and Navarro-González compared the Viking charts to those generated by their Atacama soil experiment, the results were almost identical. It appeared that there had been traces of organic carbon in the Viking sample after all.

Note that the discovery of organics does not mean life, but the organic building blocks of life. Still, this was a major finding, and for the "evidence of life" camp in the three-decade Viking life-science result discussion, welcome news. More answers should be provided by the MSL mission when it arrives at Mars in 2012.

McKay and many others have also studied the dry lake valleys in Antarctica. These odd regions, located on the coldest continent in the world, are also the very driest. Rainfall is almost unknown (true rain is estimated to have last fallen in the region about fifteen million years ago), and what moisture is deposited is rapidly depleted by the fierce katabatic (gravity-fed) winds that hurl themselves off the nearby slopes at up to 200 mph.[2] Add to this an average temperature of 4°F, and you have a rather similar environment to Mars in many important ways.

There are also lakes in this exotic region, covered with a thin sheet of ice year-round. They are highly alkaline and ten times as salty as the open sea. These traits make them somewhat akin to the water found on Mars in the distant past. But it is the dry areas that fascinate the Mars crowd. Life is highly challenged in the dry valleys; there is no vegetation and no major life-forms anywhere to be found. Life, where it can be discerned, hangs on tenaciously by a thin thread. But life finds a way, and life there is: In the lakes, mats of algae can be found. Bacteria, yeast, and varied fungi can be found in the nearby soil. The most advanced life takes the form of nematodes, tiny worms.

What fascinates people like McKay is the phenomenon of *eternal permafrost*. Large regions of the Canadian Arctic and other similar regions are made up of permafrost, but it melts in the summer season, changing chemistry and soil dynamics. The dry valleys, however, consist primarily of rock and soil over ice which never thaws—much like a lot of Mars. This has proved to be an ideal environment for a number of research projects.

Some of these projects are mechanical in nature. As the soil is so Mars-like, prototypes for drills and scoops, destined to fly on Mars probes, are tested here regularly. The sampling system for the Mars Phoenix was tested here. Peter Smith, the head of that program at the University of Arizona, was struck by the similarity between this region and the Red Planet. "Those upper valleys are the best analog for the Phoenix site," he has stated. "The soil temperatures are always well below freezing, ice is stable about fifteen inches below the surface, and the extreme conditions challenge life-forms to the maximum. This is as close as we can get to Martian conditions."[3] The wet chemistry lab for Phoenix was also put through its paces here, a valuable simulation for the harsh conditions found near the Martian poles.

The drill for the MSL rover was tested in the dry valleys; it's a percussion drill that pounds the soil as it drills into it in order the get the greatest bang for the (weight) buck possible. This kind of testing is more challenging than it sounds. Deep in the hills, far from the nearest bit of civilization, teams must operate the drills while generating their own power and carrying spare parts. Everything must be helicoptered in—and out—including human waste.

Figure 29.2. THE DRY VALLEYS: A dry-valley area in Antarctica. This soil has been etched into polygons not dissimilar to those encountered by the Mars Phoenix Lander. The dry valleys can serve as test beds for Mars-bound spacecraft components. *Image courtesy of NASA.*

And the particulars of the drill must be carefully measured: allow the bit to get too hot, and it will melt the very ice it is attempting to sample. Allow the ice being drilled to refreeze, and that's the end of the drill bit, and the sample.

Beyond mechanical testing though, is the presence of microbes and bacteria in the soil. This dirt is infused with perchlorate, much like the soil of Mars has demonstrated itself to be. Besides being a traditionally toxic blend for earthly organisms, perchlorate also acts as an antifreeze, changing the freezing point of any water that might be found. And perchlorate has also been found to support some kinds of microbial life; anything learned about that relationship can help scientists like McKay understand how life might exist—even thrive in a penurious way—on Mars.

And these colonies of microbial life—some estimated to be thousands of years old—exist well below the bleached surface. Add to this the existence of hypersaline water that can resist freezing, even in Martian temperatures, and you have an ideal laboratory for Martian life-science simulation.

Even the small lakes provide a test bed; it is theorized that in some regions of Mars, similar conditions exist where ice overlaying salty water could keep that liquid from freezing. In the dry valleys, even when it is below 0°F, the water beneath the icy surface of the lake is routinely at 32° or higher. This condition, if it exists on Mars, is one more factor in providing an environment where life could exist even today. The ice also traps gasses in the water, and as much as 300 percent more oxygen and 160 percent more nitrogen have been found in the dry valley lakes than the surrounding environment. Again, if this holds true on Mars, the possibilities for life, past or present, look better and better.

Exploring the parts of Earth that mimic, in some way, Mars is a cost-effective and achievable way to learn much about future explorations of the Red Planet. Thankfully, a few intrepid individuals have paved the way and shown us that such programs and research can pay off. Here's to the Earth-bound Martians.

CHAPTER 31

THE NEW MARTIANS

A small and select group form the core of the Mars analog explorers. More join this elite cadre every year. What they have in common is a driving curiosity, a sense of adventure, and the burning desire to experience, in some form, Mars.

ROBERT MANNING, JPL[1]

Rob Manning, whom we met during his involvement with Mars Pathfinder, is such a person. While not a pioneer in Mars analog work, he made the journey from JPL to the wilds of Washington State to experience firsthand some of the challenges that would be facing his tiny Pathfinder spacecraft once it reached Mars.

"[We] wanted to go to what we call a 'grab-bag' site on Mars, where there were lots of different kinds of rocks, that we could use the nose at the end of the rover, of the APXS, and they selected our area, which was an outflow channel just south of the equator. It's a large chaotic area, where water had popped out of the ground due to reasons which at the time were not clear, and raced down this valley, carving these catastrophic flows at the mouth of Ares Vallis. You can see, even from space, what appears to be water flow channels, and sandbars; it's fantastic."

After scrutinizing satellite photos and conferring with the geological team, an Earth analog gradually emerged for testing of the Sojourner rover.

"So we went to Moses Lake, Washington. It turns out that about fifteen thousand years ago, there was a giant lake in Missoula, Montana, which was stopped up by a giant iceberg. One day it broke, and it sent a cascade of water about fifty feet high, maybe miles across, racing across Washington, the Columbia River, and all the way back to the ocean, and down to Salem, Oregon.

"There are rocks from Montana, which you can still find, down in North Central Oregon, that were carried down there. These flowed in a catastrophic flood that scoured away the land, and of course this event was huge on Earth, but it was actually much smaller than what happened on Mars, so we did a fantastic field trip that I will never forget. We went all the way around Washington, with some of the top geologists; we talked about our rovers, airbags, rock distributions, how we could test the airbags on Pathfinder.

"We brought a little rover, it's a little rubber-wheel rover, and we tried driving on the terrain to see how it would work on different rock distributions and types. It was just like we thought the rock's density and distribution would be at our landing site on Mars. There had been a brushfire, so there was no brush around, just charred earth and dirt and rock, and it looked much like Mars except for its darker color. We also brought our lander structure to assemble on the ground to see what its orientation might be on those kinds of rocks once it set down. We wanted to get a feel for what we might be dealing with eventually. It was all very, very helpful."

The trip was a resounding success. The lessons learned from driving across the rock fields of Washington and Oregon were of great benefit when planning traverses on Mars and continued, along with other field simulations, to inform during the later missions of the Mars Exploration Rovers. And all this leads up to the granddaddy of rover mission, the Mars Science Laboratory. Rob Manning will be there. . . .

Dr. CHRIS McKAY, NASA AMES RESEARCH CENTER[2]

When peering into the annals of earthly Mars analogs, the name Chris McKay surfaces again and again—as one might expect, for he has done as much of the work as, and more than, almost anyone. How he came to this work, and this passion, is by now a familiar story.

"I was interested in physics and astronomy as an undergraduate. It was the Viking results in 1976—my first year of graduate school—that sparked my interest in Mars. Then, in 1980, I had the opportunity to be part of NASA's first group of planetary-biology interns. I worked at NASA Ames for the summer and met Imre Friedmann, who was then at Florida State University. Later that year, I went to Antarctica with him, and this cemented my interest in fieldwork related to astrobiology.

"My main interest in science is the search for a second genesis of life. Extremophiles, organisms living in extreme environments, are of interest because they show the limits of life. To explore this, we've gone to many places in the world where life is in dry or cold, or dry *and* cold. So lots of deserts, alpine, and polar regions. Different from all the others was an expedition into the Crystal Cave in Mexico, where we were testing an instrument for noninvasive organic analysis."

The trip to Crystal Cave was a harrowing one. Deep inside Mexico's Naica Mountain, the cavern is a place of razor-sharp rocks, yawning crevasses, and boiling hot water and steam. It is not a place for the faint of heart . . . but then again, neither is Mars. Sitting astride huge fault lines, the caves rest atop a huge magma chamber over a mile below. Hot metal-rich fluids circulate through the fault cracks and the cave itself. One chamber, discovered by miners in 2000, is huge, sporting enormous gypsum crystals far larger than anything found in nature and looking like something out of a Jules Verne fantasy. It is also filled with hot gasses, over 120°F, that are deadly.

Chris McKay traveled there with a team of scientists in 2007. Wielding a spectrometer, he was able to search for organic substances within the giant crystals. The results would inform research scheduled by the MSL mission launched in November 2011.

"The work in Crystal Cave was to test an instrument for noncontact detection of organics. By 'noncontact' I mean several meters away. On Mars we want to be able to point our laser at a rock and determine if it has organics. If it does, we might then go to the effort to command the rover to go get the rock and do further analysis on it."

But it is the truly extreme environments—the Atacama and the Antarctic—that continue to draw McKay away from the comforts of home: "The Atacama and the Antarctic work remain the most rewarding both in terms of science and in terms of persistent and broad interest on my part. I work in these locations longer and more persistently than anywhere else. The main goal of our work is to understand how life survives in Mars-like conditions and how evidence for life is preserved in Mars-like conditions. How this translates into activities in the field varies considerably from site to site and from year to year. One year in the Atacama we may be focused on how cyanobacteria survive in the dry limit [how dry an environment can be and still support some form of life]. The dry limit for most desert life is crossed in the Atacama Desert. Research results show that the Atacama Desert soils were 'Mars-like'; meaning that they had very low organics, very low DNA, and no culturable microorganisms, as well as the presence of a nonbiological oxidant [perchlorate]. [We have used] Atacama soils as a surrogate for Mars soil in a test of the Viking [experiments] and the effects of perchlorate. Results of this research show that the perchlorate discovered on Mars by the Phoenix mission explains the lack of detection of organics in the Viking mission."

Then the next year might be in Antarctica.

"We are focusing on the high elevations of the Antarctic dry valleys. In particular, University Valley at 1,700 meters [about

5,500 feet] and the nearby valleys. These valleys are so cold and dry that they are the only place we know of on Earth that has 'dry permafrost.' Everywhere in the Arctic and most of the Antarctic, the top of the permafrost melts in the summer, forming what is known as an 'active layer.' This is the depth to which [a] melting and wet condition penetrates in the summer. In University Valley we find ice-cemented ground that is from a few centimeters to forty centimeters deep [about 1.3 feet], and it never forms a bulk liquid phase. In simple terms, it never melts. For all intents and purposes, the only forms of [water] we have in University Valley are ice and vapor—just like Mars. Our project here is to develop a drill that can work in these Mars-like conditions and to investigate the physics and microbiology of this Mars-like dry permafrost environment."

Ice and rock drills for missions like the Mars Science Laboratory and beyond benefit from this research, and it is ongoing.

Like Indiana Jones, McKay roams the world looking for new adventures and seeking the truth. But unlike that fictional swashbuckler, his mind is forever in the skies . . . on a frosty red planet named Mars.

Dr. ROBERT ZUBRIN, PIONEER ASTRONAUTICS AND THE MARS SOCIETY[3]

Robert Zubrin has been described as a renegade and a visionary. His overriding characteristic seems to be that he tells it like it is—he says what he really thinks, not what is politically expedient or convenient. And what he is trying to tell us, those who will listen, is that it's time to head off to Mars. Long impatient with US efforts in this direction, he was a founding member of the effort to create the Flashline Station simulation of a Mars mission on Earth. He also wrote a bestselling book about a possible future for crewed Mars-exploration efforts, *The Case for Mars*, and it galvanized his efforts.

"The book was very successful, sold one hundred thousand copies, and is currently in its sixteenth printing. I got four thousand letters, and they came from all over the world—just a remarkable number of people. Astronauts, people from JPL, twelve-year-old kids from Poland, an incredible assortment of people wrote these letters. I looked at this and I talked with Chris McKay about it, and I said, look at this, look at all these people. They're all saying the same thing . . . ultimately what they're saying is: How do we make this happen? I thought if we put these people together and form an organization, we'd have a force that might be able to make it happen.

"So I used the list of the people who sent me letters as a mailing list, and we announced an annual convention of the Mars Society. That first year, seven hundred people showed up, and they came from all over the US and over the world, so we formed the Mars Society that way. We then decided we would do three forms of activity. One was general public outreach, just to spread the faith. The second would be political work, and the third, a project of our own to build a Mars-analog research station."

Such a simple thing to say, but much more difficult—and at times harrowing—to do. Money had to be raised; a location had to be selected. Once the site was decided upon, the station had to be designed and prefabricated. This accomplished, it would have to be flown—in pieces—to the Arctic and reassembled there. The story is very complex, and is wonderfully covered in Zubrin's book. Suffice it to say that the task was completed, not once but twice (the second station is in Utah), and the results speak for themselves. After years of unflagging effort, Mars now exists on Earth.

"What we're into is not so much the field science itself, but something about the exploration process. Besides the Arctic location, we have a separate station in the Utah desert . . . we take a crew, it's usually six people, and we have them do a sustained field exploration in geology or biology while operating under as many constraints as we can impose upon them. We try to find out what's going to work on Mars: what technology would work, what skill mixes would work, what character mixes, a lot of human-factor work, how you want to organize the crew. Do you have people on [military-style] watches, do you do it with everyone on the same schedule, and who leads it? Is it mission control, or the crew commander? What kind of field mobility systems are optimal, what kind of instruments are optimal—all kinds of stuff. So we now have had eleven crews at the Arctic station, and over one hundred crews [at] the desert station. At this point, over six hundred people have been part of one crew or another of our stations.

"We've had all kinds of crews: mixed international, all-German crews, all-Australian crews, we've had all-male crews, all-female crews, everything you could think of. So now some six hundred people have had some experience of what the challenge is like, and they take that knowledge with them back into the workplace or wherever they are."

When he looks into the future, Zubrin is optimistic about the ongoing role of the Mars Society. NASA is another story.

"We're going to continue [our program]; we've got ten crews lined up for this year. We're also looking to have a significant fraction of the desert station become an educational program in conjunction with some educational institutions. They could send students to them and become a program that gives out credit and has professors involved. That's another aspect of this that we will open up."

And regarding NASA?

"I think that with this extremely difficult budget environment about to descend upon the whole discretionary part of the government, the NASA program is going to have to be made much more defensible to adequately fund it. The situation is so bad that even the ExoMars robotic missions are in great peril, and currently their plan is to run a spaceflight program in the next ten years without any discernable objective. It's hard to understand how that could survive in the difficult environment we're about to face.

"I think frankly that NASA needs to formulate a program for a human Mars mission, because that is the human spaceflight mission that's worth doing. [They need to] propose this not as a ten-year trial but a project to achieve the mission, and say, 'Look, this is the kind of human spaceflight program Americans want, that they deserve. They really want a space program that's actually going somewhere, and to say we can't afford any kind of space program is in any account a decline."

And that's the truth, as seen by a visionary. We would be wise to listen.

CHAPTER 32

THE ROAD AHEAD

SHACKLETON BASE, CHRYSE PLANITIA, MARS
REPORT FSC-17785.88
(Personal entry)
SOL 2344

Dr. J Carter, MSPI

It's Thursday the 17th of May Prime*, 2029. What light comes through the window is dun-colored and dim, which matters little because the glass is so sand-abraded that you can't see much anyway.

We're supposed to finish the installation of Module 6 this month, but this dust storm throws our schedule off a bit. It's a global sandstorm that has grounded all interbase flights, and the Japanese crew will be stuck in orbit for at least a week, so if the Hab Module is a bit late, they won't be here to notice.

It will be interesting having the Japanese here. The US missions first arrived in 2026, with the Chinese joining us on the surface six months later. Phobos base is now just a way station. The two colonies operated separately, until the US base was hit by a small meteorite and became unlivable for a few months; we had to shelter at Guan-Yin Station until repairs were accomplished. Once we and the Taikonauts got better acquainted, it was (wisely) decided to join the two bases with a tunnel. The food got a lot better after that . . .

Too bad the Russians bugged out three months earlier. The Beijing duck could have gone well with a good vodka! Maybe when the economic situation back home improves, they will return.

So now we'll have Japanese nationals blending into the mix. There are already four nationalities represented on Mars, but this will be only the third official habitation built. No matter; I like hydroponic wasabi too.

We just received word that the windstorm is abating, which is good news after three weeks of howling noise! While high winds are not as nasty as they are on Earth (the low pressure and all), the sand still gets into everything, so we'll have a lot of work to do. We will also have to do a windblown-perchlorate scrub of the airlocks and service areas—that's a ton of work. Then perhaps next week I can take that hike out to Viking Park I've wanted to do, and get a look at the old spacecraft. Hard to imagine that it landed here, big dumb and blind, over fifty years ago!

Signed: Julia Carter
Sr. Atmospheric Scientist
US Mars Program

NOTE: As the Martian year is close to two Earth years, the months have been doubled for calendrical convenience. May Prime (May) is the second iteration of May in a Martian year.*

This is science fiction for now, but not an unlikely scenario for the future. The path currently being traveled by the world's spacefaring powers may well lead to multiple settlements on the moon, and even Mars, within the next thirty years. And while the United States and Russia are likely to be first among them, Asia is catching up fast and moving ahead with great determination. The future of Mars exploration—by someone—seems assured.

A crewed mission to Mars is an enormous undertaking. Planners have been envisioning such a mission since the 1950s. Wernher von Braun, the father of the Saturn V moon rocket, famously laid out his plans for flights to Mars in both *Collier's Weekly* magazine in the 1950s and later on television courtesy of *The Wonderful World of Disney*. His vision, vast and optimistic, blazed in the minds of children and adults alike for years.

After the Apollo lunar missions in the 1970s, both US aerospace and NASA had conducted copious studies about the use of Apollo-era hardware for a manned Mars flight. Many thought that it could be accomplished by the early 1980s . . . and they may have been right, but as it so often does, fate had other plans.

The last of the Apollo hardware, scavenged from the final three (canceled) lunar landing missions of Apollo 18, 19, and 20, was used for Project Skylab in 1973 and the Apollo-Soyuz Test Project in 1975. After that, the bulk of NASA funding was transferred to building and operating the Space Shuttle. The remaining Saturn Vs became the world's most expensive museum exhibits. Mars waited . . . and waited.

Plans came and went, but none were granted the go-ahead for manned exploration. Private groups joined the discussion, but the best they could hope to accomplish was to build public sentiment. Unlike Earth orbit and, perhaps, the moon, Mars is beyond the reach of private enterprise for the foreseeable future.

Then, in 2004, President George Bush declared Mars a national goal. Not soon, for a return to the moon would come first. But the mandate for future crewed space vehicles would include designs ultimately capable of the long voyage to Mars. Planned by NASA as a replacement for the Space Shuttle, this mission was canceled by the Obama administration in 2010, after an expenditure of about $8 billion. The revised program may look something like this:

2010–2015

ORION (UNITED STATES): NASA's first post-shuttle program includes the Orion Multi-Purpose Crew Vehicle (MPCV) and unspecified boosters; as of now, only the capsule is under active development. Tests of a new booster based on the shuttle's solid booster have been canceled, but others are still on the drawing boards. The most current is the Space Launch System, a Saturn V–class heavy booster that would be capable of hurling the Orion capsule and associated Mars-class hardware out of Earth's gravity well. This capability would be the first time since the mid-1970s that we would be able to leave Earth orbit.

SPACEX (UNITED STATES, Private): Space Exploration Technologies, known as SpaceX, is the brainchild of Elon Musk of PayPal fame. This private enterprise is already completing work on the Dragon® capsule, slated to provide resupply missions to the International Space Station and, eventually, manned access. Future plans for SpaceX include a heavy-lift booster, utilizing a cluster of engines for its tremendous lifting capability. Also in the general realm of the Saturn V, this rocket would most likely be ready far earlier than NASA's new booster and at a fraction of the cost. And SpaceX has specifically set Mars as its goal.

SOYUZ (RUSSIA): While the Russian Soyuz capsule is seen today primarily as a shuttle vehicle between Earth and the International Space Station, it should be remembered that this very capable spacecraft was born of the space race and was designed to travel to the moon and back. In fact, it was successfully tested in unmanned flights around the moon. Its current configuration could be uprated for longer-duration travel, possibly even as a part of a Mars-bound complex.

2016–2020

NASA (UNITED STATES): Little is certain at NASA beyond the development of the Orion capsule. With luck, there will be a crew-rated booster available to fly this capsule to lunar orbit and beyond. The Space Launch System, or SLS, is a large booster currently under development, but its ultimate fate is uncertain.

SHENZHOU/SOYUZ (CHINA): The Chinese space agency has licensed the venerated Soyuz design from Russia and made vast improvements within. This more modern design has flown Chinese crews into Earth orbit three times (as of press time), and could also be augmented to travel to the moon and beyond. China has made clear its intention to land crews on the moon before 2020, and Mars may not be far behind.

2021–2030

NEAR-EARTH OBJECT (NEO) MISSION (UNITED STATES): Current planning around this manned mission is still in the planning stages, but a voyage out to a large asteroid is currently on the drawing boards. Useful both scientifically and strategically (NEOs represent a significant threat to Earth), this project could be accomplished far easier than a Mars landing, due primarily to the lack of a need to enter and escape the Martian gravity well.

CONSTELLATION/MARS (UNITED STATES): By 2030, NASA may be heading off to Mars on crewed missions using either Orion- or Dragon-type capsules mated with habitation and propulsion units. Little is known at this point what form such a spacecraft might take, but it will likely be used as the basis of a transit vessel and a Mars-lander/ascent-vehicle design.

PHOBOS MISSION (UNITED STATES and/or ESA/ RUSSIA): There is growing sentiment that before a crewed mission to Mars itself is attempted, a small station might be set up on the larger of the two Martian moons, Phobos. This would allow for close-in observation of Mars without the added complication (and launch mass) of a crew module capable of atmospheric entry, landing, and ascent. Operating on Phobos would be only slightly more difficult than landing on Earth's own moon, a feat we accomplished forty years ago. Such a mission would further serve as a demonstration of the ability to make the long voyage between Earth and Mars, which is bound to be quite taxing on both crew and spacecraft.

The prospects for missions to the Red Planet have never been great. Always, these long-term, big-vision programs are subject to delays, cancellations, and political wrangling. Indeed, since the days of the Kennedy moon challenge and the space race, there have been few space programs that flew as initially planned, and fewer still that have reached fruition without massive cost cutting and mission shrinkage. Even the now-discontinued space shuttle was a mere shadow of NASA's original vision.

Mars beckons to all humanity. The United States hears the call, as do Russia, China, India, Japan, and Europe. Someone will go at some point. Who goes matters less than the fact that we do go, because without some kind of new goal, the spirit of exploration and "reaching out" may well leave our species.

As the nearest planet to ours, and the only likely candidate for colonization in our solar system, Mars is the next logical step. As Ernest Shackleton, the famed British polar explorer, once said: "Optimism is true moral courage . . . difficulties are just things to overcome, after all."

Who among us, which nations among the ever-growing cadre headed for space, will summon the courage and fortitude to go?

NOTES

CHAPTER 1. THE FIRST MARTIAN

1. The Soviet unmanned Mars program was spectacular in its persistence and its failures. While successful with other planets, notably Venus, Soviet-era Mars missions were notorious in their failure rate. Two years prior to the successful landing of Viking 1, in August 1974, the Soviet Mars 6 and 7 entered Martian space. Mars 7 failed before descending, but Mars 6 actually "landed" on the surface of Mars, transmitting a few minutes of data, mostly unintelligible, just prior to touchdown. Since the fall of the Soviet Union, and with increased cooperation with the European Space Agency, the program has seen some shared success. Yet as recently as November 2011, the Russian-led Phobos-Grunt sample return mission to the Martian moon Phobos failed after launch.

2. Among other things, the lander was baked in swirling clouds of nitrogen gas for over forty hours. The goal was to make sure that no earthly organisms polluted either Mars or the life science experiments. In fact, NASA/JPL had devised an entire program of "planetary quarantine" leading up to this mission. Ironically, in the intervening decades, it has become clear that the Martian environment is so very toxic, with high levels of solar radiation and powerful oxidizing agents present in the soil, that most anything that could have hitched a ride on the lander would have been dead shortly after touchdown.

CHAPTER 2. MARS 101

1. The "Goldilocks Zone," also known as the "habitable zone," is a unique combination of circumstances that must combine (it is thought) for a planet to be able to sustain life. These include: a proper distance from the star it orbits to be able to sustain liquid water on the surface, a size generally similar to Earth's, a star that is not hostile to life-supporting conditions, and a position within the larger galaxy that is not hostile to life (i.e., does not have radiation levels that are deadly to life-forms). This does not necessarily imply that the planet itself can support life. The environment needed to sustain carbon-based life-forms is a different set of conditions and variables. (The name of this zone comes, of course, from "Goldilocks and the Three Bears," in which Goldilocks tastes three bowls of porridge and rejects two for being too hot or too cold, but determines that the third is just right.)

2. In the early 1960s, mascons were first encountered by the early unmanned orbiters, whose orbital paths were affected in unexpected ways over certain regions of the moon. They were of much concern to those planning the descent paths of the Lunar Modules, and that was one reason for the "barnstorming" flight of Apollo 10, in which astronauts approached but did not land on the lunar surface. They wished to further explore the effects of these anomalies on descent trajectories.

3. The "terrestrial" planets in our solar system are, in order from the sun, Mercury, Venus, Earth, and Mars. They are characterized by rocky composition, a solid surface, and roughly similar sizes.

4. Mars suffers from what is termed *low thermal inertia*, which means that the surface heats quickly in sunlight. There are no oceans present to dampen these effects with clouds or their own weather systems. Martian versions of trade winds can cycle around the planet at very high velocities, though in the thin

atmosphere their effects are not the same as they would be on Earth. Wind velocities of over 100 mph are hypothesized. A 30–50 mph blow can cause dust to lift, often for weeks at a time; higher velocities lift more dust. The chances of a planetwide dust storm in a given year seem to be about one in three.

5. One such body of theory is called *panspermia*. It proposes that life can survive dormant for long periods in space, hitching a ride aboard a meteor or an asteroid. If the transporting body then impacts another planet, and the conditions are right, it could return to an active state and begin to evolve and adapt. Some propose that life started on Mars and was transported to Earth. And while no direct evidence has yet surfaced, meteors of both lunar and Martian origins have been found on Earth.

CHAPTER 3. IN THE BEGINNING: A SHINING RED EYE

1. Anonymous source from Middle Ages Europe (ca. approx. 1400 CE), in Willy Ley, *Mariner 4 to Mars* (New York: Signet, 1966).

2. Retrograde motion can be thought of this way: imagine that you are driving on a racetrack on the inside lane. Mars is driving on the outside lane. For most of the lap, if you take your eyes off what's in front of you and look over at Mars, the background is moving the same direction relative to the planet. However, as you complete your lap (you are moving faster than Mars, about twice as fast), and you approach and pass Mars, it seems to move in the *opposite* direction for a brief period. Now picture Earth's orbit and Mars's orbit outside of it—a similar situation applies. Retrograde motion of Mars appears every two years.

3. English astronomer, 1860.

4. Camille Flammarion, *Popular Science* 4, no. 9 (December 1873): 189. English translation from the French.

5. G. V. Schiaparelli, *Osservazioni astronomiche e fisiche*

sull'asse di rotazione e sulla topografia del pianeta Marte, vol. 4 (Rome, Italy: Coi Tipi del Salviucci, 1896).

6. G. V. Schiaparelli, "Schiaparelli on Mars," *Nature* 51 (November 22, 1894): 89.

7. John Michels, "Review of 'Mars' by Percival Lowell," *Science* 4, no. 86 (August 21, 1896): 233.

8. Percival Lowell, *Mars* (Boston: Houghton, Mifflin, 1895).

9. Ibid.

CHAPTER 4. THE END OF AN EMPIRE: MARINER 4

1. The number of unmanned explorations sent to Mars is nearing forty, yet almost half have been failures. Of these, the vast majority were from the Soviet Union. While successful with many of their missions to Venus, Russian plans for Mars exploration have yielded little success.

CHAPTER 5. DR. ROBERT LEIGHTON: THE EYES OF MARINER 4

1. Dr. Robert Leighton, interview by David DeVorkin, August 5, 1977, Niels Bohr Library and Archives, American Institute of Physics, College Park, Maryland, http://www.aip.org/history/ohilist/4738_1.html (accessed July 2011).

CHAPTER 7. DR. BRUCE MURRAY: IT'S ALL ABOUT THE IMAGE

1. Dr. Bruce Murray, interview by Rachel Prud'homme, March 1984, courtesy of the Caltech Archives, the California Institute of Technology.

CHAPTER 8. AEOLIAN ARMAGEDDON: MARINER 9

1. Mariner 9 would be the first of JPL's Mars missions to set the high benchmark to which all now seem to be held. Its primary mission was set at ninety days, just two months longer than the dust storm raged. But the spacecraft sent back images and data for almost a year, extending the mission duration by a factor of four.

CHAPTER 9. DR. LAURENCE SODERBLOM: THE EYES OF MARINER 9

1. Dr. Laurence Soderblom, interview by the author, August 2011.

CHAPTER 10. VIKING'S SEARCH FOR LIFE: WHERE ARE THE MICROBES?

1. The Soviet Mars missions for the 1973 opposition were Mars 4, 5, 6, and 7. Mars 4 and 5 were orbiters; Mars 6 and 7 were landers. Mars 4 made its way to the planet, but an error in the computer allowed it to flyby the planet as earlier craft had (by design), and the images it returned were a repeat of previous missions. Mars 5 made it to Mars but failed in orbit after less than ten days, returning some data. Mars 6 was a lander, and apparently made it to the surface, albeit at a higher rate of speed than intended. It transmitted data for a few minutes, but the onboard computer seemed to have suffered degradation during the flight and the data returned were unusable. Mars 7 was another lander, but it separated from its carrier spacecraft about four hours early and missed the planet altogether. These four failures must have been even more heartbreaking than most, as they represented a huge investment in time and resources for the Soviet unmanned program. The loss in national prestige cannot be overestimated.

2. *Utopia Planitia*—the "Nowhere Plain"—sounds like an odd translation to the modern ear. But the translation from the Greek is: *oi* ("not") and *topos* ("place") equating "nowhere." So the modern association of a perfect society does not apply in this case.

3. The Viking sampler arm was an ingenious design. Rather than carry a heavy, pipelike arm (as later landers have indeed done), the Viking's arms were carried on a spool. It was designed like two giant metal tape measures affixed front-to-front to create an elliptical profile. As the flattened metal tape rolled off the spool, it sprang into the metal's memorized shape and became rigid. It could extend for over ten feet in this fashion and was strong enough to hoist small soil samples upward to the sample containers onboard. This also gave the arm an almost infinite sampling range across its entire length.

CHAPTER 11. DR. NORMAN HOROWITZ: LOOKING FOR LIFE

1. Dr. Norman Horowitz, interview by Rachel Prod'homme, July 1984, courtesy of the Caltech Archives, the California Institute of Technology.

CHAPTER 12. RETURN TO MARS: MARS GLOBAL SURVEYOR

1. After the loss of the space shuttle *Challenger* in 1986, spacecraft that required an upper stage boost to depart Earth orbit were, in general, rerouted to expendable rockets such as the Delta, Atlas, or Titan.

2. Hematite is a mineral that we will encounter in our Mars discussions again and again. It is a type of iron oxide, Fe_2O_3. It is harder than iron but more brittle. Important for Mars explorers,

it is often found in areas that once hosted bodies of standing water, and it can condense out of water. It collects on the bottom of lakes and ponds, and it also can be found near hot springs. Alternatively, it can be found as a result of volcanic activity.

3. Great noises were made about the "Face on Mars" by some. While generally dismissed by the scientific community upon "discovery," many wanted to believe—or hope—that it represented a message from an advanced civilization. Even when the improved images came in from MGS, some continued to support this belief. A few have made this into a cottage industry, and significant profits have resulted. This is currently a fringe industry at best.

4. In the words of JPL's internal review:

> Mars Global Surveyor last communicated with Earth on Nov. 2, 2006. Within 11 hours, depleted batteries likely left the spacecraft unable to control its orientation.
>
> "The loss of the spacecraft was the result of a series of events linked to a computer error made five months before the likely battery failure," said board Chairperson Dolly Perkins, deputy director-technical of NASA Goddard Space Flight Center, Greenbelt, Md.
>
> On Nov. 2, after the spacecraft was ordered to perform a routine adjustment of its solar panels, the spacecraft reported a series of alarms, but indicated that it had stabilized. That was its final transmission. Subsequently, the spacecraft reoriented to an angle that exposed one of two batteries carried on the spacecraft to direct sunlight. This caused the battery to overheat and ultimately led to the depletion of both batteries. Incorrect antenna pointing prevented the orbiter from telling controllers its status, and its programmed safety response did not include making sure the spacecraft orientation was thermally safe.
>
> The board also concluded that the Mars Global Surveyor team followed existing procedures, but that proce-

dures were insufficient to catch the errors that occurred. The board is finalizing recommendations to apply to other missions, such as conducting more thorough reviews of all non-routine changes to stored data before they are uploaded and to evaluate spacecraft contingency modes for risks of overheating.

"We are making an end-to-end review of all our missions to be sure that we apply the lessons learned from Mars Global Surveyor to all our ongoing missions," said Fuk Li, Mars Exploration Program manager at NASA's Jet Propulsion Laboratory, Pasadena, Calif.

Jet Propulsion Laboratory, news release no. 2007-040, JPL public affairs office.

CHAPTER 13. ROBERT BROOKS: IT TAKES A TEAM, MARS GLOBAL SURVEYOR

1. Robert Brooks, interview by the author, September 2011.

CHAPTER 14. ROVING MARS: SOJOURNER, THE PATHFINDER

1. From JPL's robotics section report on Pathfinder:

The Mobility and Robotic Systems section led the development of both software and electronics for the Sojourner rover. Software enabled autonomous control, sensing, and communication. Onboard autonomy consisted of simple behaviors for navigation, based on commanded objectives along with sensed terrain and vehicle position/orientation. Terrain sensing was performed with cameras

> and laser striping, while Sojourner's position and orientation were measured by wheel odometry, accelerometers, and a z-axis angular-rate sensor. The onboard processor was a flight-qualified Intel 8085 running at 100 KIPS, and all software was written in C.
>
> In addition to the onboard control software, section personnel developed the ground control software for the rover, and provided operations expertise during the 83-day mission. The rover visited 16 science locations, traversed more than 100 meters, captured more than 500 images, and nearly circumnavigated the lander. Due to the use of lander-based operations and the roughness of the terrain, all travel was effectively restricted to line-of-sight locations not greater than 12 meters from the lander. In addition to specification of targets and drive paths, regular corrections of vehicle position and attitude were provided by operators based on the lander imagery.

Jet Propulsion Laboratory, "Flight Projects—Pathfinder: Mars Pathfinder Rover: Sojourner," Robotics, http://www-robotics.jpl.nasa.gov/projects/PATH.cfm?Project=4 (accessed September 2011).

CHAPTER 15. ROBERT MANNING, MARS PATHFINDER: BOUNCING TO MARS

1. Robert Manning, interview by the author, September 2011.

CHAPTER 16. MARS EXPRESS: ON THE FAST TRACK

1. As has been mentioned elsewhere in this book, the answer is not quite so simple. Of the many missions the Russians have dispatched to Mars (both as the Union of Soviet Socialist Republics and as the Russian Federation), a few have enjoyed partial success. In 1962, two years before Mariner 4, the Soviet Mars 1 relayed back some data but failed well before reaching Mars (it was a flyby mission). In 1971, Mars 2 (there had been many in between Mars 1 and 2, just with different naming schemes) successfully orbited Mars (just after Mariner 9 did the same). The craft operated and sent back data, but the Martian dust storm of that year made the data unrewarding. The lander piggybacked onto the mission crashed. Later in 1971, Mars 3 landed on the planet but operated on the surface for less than one minute. Mars 5 in 1973 entered orbit but failed shortly thereafter. Mars 6, a lander, sent back data on its way down to the surface but failed before landing. And so forth. The Russians have had much better luck with the exploration of Venus.

2. Mars 96 failed too. The Russian launch vehicle let the mission down; the fourth stage failed to reignite for a second burn.

3. For further information, see V. Formisano, S. Atreya, T. Encrenaz, N. Ignatiev, and M. Giuranna, "Detection of Methane in the Atmosphere of Mars," *Science* 306, no. 5702 (2004): 1758–61.

4. As this book goes to press, NASA has just announced that, due to budget cuts, it will no longer be cooperating in the ExoMars mission.

CHAPTER 17. A LAUGH IN THE DARKNESS: THE GREAT GALACTIC GHOUL

1. "Uncovering the Secrets of Mars" (interview with Donna Shirley, former manager, JPL Mars Exploration Program), *Time*, July 4, 1997.

2. NASA/JPL online historical document, http://mars.jpl.nasa.gov/programmissions/missions/log/ (accessed January 27, 2012).

3. Jet Propulsion Laboratory Media Relations Office, press release, September 23, 1999, http://mars.jpl.nasa.gov/msp98/news/mco990923.html (accessed January 27, 2012).

4. Ibid.

5. Jet Propulsion Laboratory Media Relations Office, press release no. 99-134, November 10, 1999.

6. Jet Propulsion Laboratory Media Relations Office, press release no. 01-52, March 26, 2001.

7. Public Broadcasting Service, "NASA in Question," Online *NewsHour*, April 14, 2000, http://www.pbs.org/newshour/bb/science/jan-june00/nasa_4-14.html (accessed January 30, 2012).

CHAPTER 18. 2001: A MARS ODYSSEY

1. Or perhaps their own playbook. Up through Viking and until Mars Pathfinder and Mars Global Surveyor, most Mars-bound missions had been launched in pairs. When one US mission failed, as often occurred, the other always succeeded. The same conceit could be applied to systems internal to the spacecraft: hardware backups and twin units for redundancy.

CHAPTER 19. Dr. JEFFREY PLAUT: FOLLOW THE WATER

1. Dr. Jeffrey Plaut, interview by the author, August 2011.

CHAPTER 20. TWINS OF MARS: SPIRIT AND OPPORTUNITY

1. "Sojourner's "rocker-bogie" mobility system was modified (from Pathfinder) for the Mars Exploration Rover Mission. To account for the extreme difference in weight and center of gravity from Sojourner, the mobility system on the Mars Exploration Rovers is in the back of the vehicle. The wheels are, naturally, larger and have evolved in design. Each wheel is approximately ten inches in diameter and has a unique spiral flecture pattern that connects the external part of the wheel with the spoke to absorb shock and prevent it from transferring to other parts of the rover. The rocker-bogie design allows the rover to go over obstacles (such as rocks) or through holes that are more than a wheel-diameter in size. Each wheel also has cleats, providing grip for climbing in soft sand and scrambling over rocks." Jet Propulsion Laboratory, "In-situ Exploration and Sample Return: Autonomous Planetary Mobility," http://marsrovers.nasa.gov/technology/is_autonomous_mobility.html (accessed September 2011).

CHAPTER 21. Dr. STEVE SQUYRES AND THE MARS EXPLORATION ROVERS: DREAMS OF ICE AND SAND

1. Dr. Steven Squyres, interview by the author, August 2011.

CHAPTER 22. MARS IN HD: MARS RECONNAISSANCE ORBITER

1. For the sake of comparison: Deep Space 1 (comets) = 15 gigabits of data return, Mars Odyssey = 1012 gigabits, Mars Global Surveyor = 1759 gigabits, Cassini (Saturn) = 2550 gigabits, Magellan (Venus) = 3740 gigabits, and Mars Reconnaissance Orbiter = 34 terabits, or well over three times the others combined.

CHAPTER 23. DR. RICHARD ZUREK, MRO: I CAN SEE CLEARLY NOW . . .

1. Dr. Richard Zurek, interview by the author, August 2011.

CHAPTER 24. TWINS OF MARS: SPIRIT AND OPPORTUNITY, PART 2

1. The Antarctic volcano was, in turn, named after the HMS *Erebus*, an Arctic exploration vessel belonging to the British Royal Navy. And finally, this ship was named after the mythological dark regions of Hades (or hell) in Greek mythology.

CHAPTER 25. FROM THE ASHES, LIKE A PHOENIX

1. This was the first time such an arrangement had been tried to this degree. Plenty of missions had been run in cooperation with universities, but never had the actual mission control room been remotely located (landing was, as usual, handled at JPL). It was new territory—for both NASA/JPL and the University of Arizona. The university had to cobble together a control center, complete with high-speed internet connectivity, backups for both this

and power supplies, and so forth. In the final analysis, it worked to the general satisfaction of the parties involved.

2. A commercial equivalent version of this chip came onto the market about 1991 in Macintosh® computers. But given the efficiency of the programming it was sufficient power for the missions it powered.

3. The Thermal and Evolved Gas Analyzer (TEGA) units work by receiving a sample of soil, sealing the covers, and baking the soil by slowly raising the temperature at a constant rate. The power used for heating is carefully tracked. The method is called *scanning calorimetry* and shows transitions from solid to liquid to gas of the components in the sample. At about 1800 degrees the ice in the sample vaporizes. This vapor travels to a mass spectrometer, which measures mass and concentrations at a molecular level. Organic molecules may be detected via this method.

CHAPTER 26. PETER SMITH: POLAR EXPLORER

1. Dr. Peter Smith, interview by the author, August 2011.

CHAPTER 27. MARS SCIENCE LABORATORY: BIGGER IS BETTER

1. This is one of NASA's last RTGs (Radioisotope Thermoelectric Generators). The space agency is running low on the fuel element, usually plutonium 238. This is one of the most toxic substances known to man and has been made in only very small quantities over the years. It was a nasty byproduct of the Cold War, and the two available stocks of the material, the United States and Russia, are just about out of it. When the United States ran low, it started buying it from Russia. But MSL will use a large chunk of the dwindling supply, and there is much hand-wringing

in appropriate quarters about what will be done next. Any mission beyond Mars, and even some on that planet (as with MSL) require an RTG unit rather than solar panels. It's hard and expensive to make, and NASA does not have the funds to do so. Talks are under way with the Department of Energy to try to find a solution, but at this point, no answer is in sight.

2. Accelerometers have been used in rockets and spacecraft since the beginning of the space age for obvious reasons. Early on, their critical mission was to measure the acceleration of the rocket. Later they measured motion along other axes. And speaking of the iPhone®, in 2011, two of Apple's technological miracles were flown to the International Space Station and utilized for various experiments using, of course, a specially designed app.

CHAPTER 28. DR. JOY CRISP, MARS SCIENCE LABORATORY: DIG THIS

1. Dr. Joy Crisp, interview by the author, September 2011.

CHAPTER 29. JPL 2020: THE ONCE AND FUTURE MARS

1. ExoMars is being led by the European Space Agency with collaboration with NASA and there will probably be others. It has been through a few iterations, but at its core is an orbiter that will search for trace gases is the atmosphere and a lander for somewhere around 2016, and a possible rover for 2018. The trace-gas orbiter is searching primarily for methane, as the Mars Express mission has detected the gas in the Martian atmosphere, indicating geological activity and, possibly, life.

2. The official reason for the cancellation of the Scout pro-

gram was that as operations at Mars were increasingly focused on the surface of the planet, the $500 million budget cap for Scout missions would not be able to fund a reliable lander.

CHAPTER 30. MARS ON EARTH

1. Irene Klotz, "Viking Found Organics on Mars, Experiment Confirms," *Discovery News*, January 4, 2011.

2. Also called *drainage wind*, these strong winds occur when a higher-density air mass flows down a hill or slope to a lower-density area. They can range from a few miles per hour to hurricane force. As they descend and compress, heating can occur as the mass is concentrated. In the Antarctic, the air stays cold.

3. "Antarctic Expedition Prepared Researchers for Mars Project," NASA/JPL press release, February 2009, http://www.jpl.nasa.gov/news/features.cfm?feature=2017 (accessed October 2011).

CHAPTER 31. THE NEW MARTIANS

1. Robert Manning, interview by the author, September 2011.
2. Dr. Chris McKay, interview by the author, October 2011.
3. Dr. Robert Zubrin, interview by the author, October 2011.

BIBLIOGRAPHY OF PRINT SOURCES

Ezell, Edward Clinton, and Linda Neumann Ezell. *On Mars*. New York: Dover, 2009.

Godwin, Robert. *Mars: The NASA Mission Reports*. Vols. 1–2. Ontario: Apogee Books, 2000.

Kessler, Andrew. *Martian Summer*. New York: Pegasus Books, 2011.

Lowell, Percival. *Mars*. Cheshire, England: New Line Publishing, 2009. First published in 1897, in New York, by Houghton Mifflin.

——. *Mars as the Abode of Life*. Boston: Adamant Media, 2002. First published in 1898, in London, by Smith, Elder.

Maimone, Mark, P. Charles Leger, and Jeffrey Biesiadecki. *Overview of the Mars Exploration Rovers' Autonomous Mobility and Vision Capabilities*. Washington, DC: National Aeronautics and Space Administration (NASA)/Jet Propulsion Laboratory (JPL), 2007.

Morton, Oliver. *Mapping Mars*. New York: Picador/MacMillan, 2002.

National Aeronautics and Space Administration (NASA)/Jet Propulsion Laboratory (JPL). "JPL Technical Memorandum no. 33-229." In *To Mars: The Odyssey of Mariner IV*. Washington, DC: NASA/Caltech, 1965.

——. *Mars Climate Orbiter Press Kit*. Washington, DC: NASA/JPL, 1999.

——. *Mars Global Surveyor (MGS) Loss of Contact*. Washington, DC: NASA/JPL, 2007.

——. *Mars Reconnaissance Orbiter Press Kit*. Washington, DC: NASA/JPL, 2006.

——. *Mars Science Laboratory Radiological Contingency Planning*. Washington, DC: NASA/JPL, 2010.

——. *NASA Facts: Mars Global Surveyor*. Washington, DC: NASA/JPL, 2000.

——. *NASA Facts: Mars Pathfinder, 5-99AS*. Washington, DC: NASA/JPL, 1999.

——. *NASA Facts: Mars Reconnaissance Orbiter*. Washington, DC: NASA/JPL, 2006.

——. *NASA Facts: Mars Science Laboratory 2011*. Washington, DC: NASA/JPL, 2011.

——. *NASA Facts: The Viking Mission*. Washington, DC: NASA/JPL, 1975.

——. *NASA Facts 2001: Mars Odyssey 2003*. Washington, DC: NASA/JPL, 2003.

——. "Release 71-215." In *Mariner 9 Press Kit*. Washington, DC: NASA/JPL, 1971.

Nicks, Oran. *A Review of Mariner 4 Results, NASA SP-130*. Washington, DC: National Aeronautics and Space Administration (NASA)/Jet Propulsion Laboratory (JPL), 1967.

Turner, Martin. *Expedition Mars*. New York: Springer-Praxis, 2004.

Wilson, James. *Two Over Mars: Mariner 6 and Mariner 7*. Washington, DC: National Aeronautics and Space Administration (NASA)/Jet Propulsion Laboratory (JPL), 1969.

Zubrin, Robert. *The Case for Mars*. New York: Touchstone, 1996.

——. *Mars on Earth*. New York: Tarcher, 2003.

BIBLIOGRAPHY OF INTERNET SOURCES

SELECTED REFERENCES

"Antarctic Expedition Prepared Researchers for Mars Project." *ScienceDaily*. February 5, 2009. Accessed 2011. http://www.sciencedaily.com/releases/2009/02/090205141509.htm.

Burton, Kathleen. "Mars-Like Atacama Desert Could Explain Viking 'No Life' Results." Release 03-87AR. NASA, Ames Research Center. November 7, 2003. http://www.nasa.gov/centers/ames/news/releases/2003/03_87AR.html.

Cornell University. "Athena: Mars Exploration Rovers." Accessed 2011. http://athena.cornell.edu/.

ESA. "Mars Express." Accessed 2011. http://www.esa.int/SPECIALS/Mars_Express/index.html.

Klein, Harold, Norman Horowitz, Gilbert Levin, et al. "The Viking Biological Investigation: Preliminary Results." *Science* 194, no. 4260 (October 1976): 99–105. Accessed 2011. http://www.sciencemag.org/content/194/4260/99.abstract. doi: 10.1126/science.194.4260.99.

Mars Daily. "Digging Deep: An Interview with Chris McKay." August 15, 2006. Accessed 2011. http://www.marsdaily.com/reports/Digging_Deep_An_Interview_With_Chris_Mckay_999.html.

Mars Institute. "Haughton-Mars Project: August 2010." Accessed 2011. http://www.marsonearth.org/.

National Aeronautics and Space Administration (NASA). "A

Chronology of Mars Exploration." Accessed 2011. http://history.nasa.gov/marschro.htm.
——. "The Mariner Mars Missions." NASA/NSSDC, Goddard Spaceflight Center. Last modified 2005. Accessed 2011. http://nssdc.gsfc.nasa.gov/planetary/mars/mariner.html.
——. "Mars Pathfinder." Accessed 2011. http://www.nasa.gov/mission_pages/mars-pathfinder/.
——. "Phoenix Mars Lander." Accessed 2011. http://www.nasa.gov/mission_pages/phoenix/main/index.html.
——. *Review of NASA's Planned Space Program*. Washington, DC: National Academy Press, 1996. Accessed 2011. http://books.google.com/books?id=3z0rAAAAYAAJ&pg=PA6&dq=nasa+mariner&hl=en&ei=OOXnTr7sHozMiQKbyeH-Bg&sa=X&oi=book_result&ct=result&resnum=3&ved=0CEMQ6AEwAjgK#v=onepage&q=nasa%20mariner&f=false.
——. "Slam-Dunk Sign of Ancient Water on Mars." NASA Science: Science News. December 8, 2011. Accessed 2011. http://science.nasa.gov/science-news/science-at-nasa/2011/08dec_slamdunk/.
——. "Viking, Mission to Mars." Accessed 2011. http://www.nasa.gov/mission_pages/viking/.
——. "Viking Project Launch and Mission Operations Status Bulletin No. 7." NASA, Langley Research Center. June 23, 1975. Accessed 2011. http://www.scribd.com/doc/44042526/Viking-Lander-Sterilization.
National Aeronautics and Space Administration (NASA)/Goddard Space Flight Center. "Lunar and Planetary Science: Mars." Accessed 2011. http://nssdc.gsfc.nasa.gov/planetary/planets/marspage.html.
National Aeronautics and Space Administration (NASA)/Jet Propulsion Laboratory (JPL). "Mars Climate Orbiter Failure Board Releases Report, Numerous NASA Actions Underway in Response." November 10, 1999. Accessed 2011. http://mars.jpl.nasa.gov/msp98/news/mco991110.html.

——. "Mars Exploration Program: Historical Log." Accessed 2011. http://mars.jpl.nasa.gov/programmissions/missions/log/.

——. "Mars Exploration Rovers Press Releases." Accessed 2011. http://marsrovers.jpl.nasa.gov/newsroom/pressreleases/index.html.

——. "Mars Pathfinder." Accessed 2011. http://mars.jpl.nasa.gov/MPF/.

——. "Mars Reconnaissance Orbiter." Accessed 2011. http://marsprogram.jpl.nasa.gov/mro/.

——. "Mars Science Laboratory." Accessed 2011. http://marsprogram.jpl.nasa.gov/msl/mission/.

——. "Report from Mars, 1964–1965." Accessed 2011. http://www.scribd.com/doc/44038607/Report-From-Mars-Mariner-4-1964-1965.

——. "Spirit and Opportunity, Mars Exploration Rovers." Accessed 2011. http://www.nasa.gov/mission_pages/mer/index.html.

——. "U.S. Participation in Europe's Mars Express." Accessed 2011. http://www.http://mars.jpl.nasa.gov/express/.

National Aeronautics and Space Administration (NASA)/Jet Propulsion Laboratory (JPL)/ Planetary Data System. "Viking 1 & 2." Accessed 2011. http://pds.jpl.nasa.gov/planets/welcome/viking.htm.

Planetary Society. "Mars Exploration Rovers Update" (miscellaneous dates). Accessed 2011. http://www.planetary.org/news/2011/0901_Mars_Exploration_Rover_Update.html.

——. "Mars Express." Accessed 2011. http://planetary.org/explore/topics/mars_express/.

"Report: Mars Pathfinder," *Science* 278, no. 5344 (December 1997): 1734–42. Accessed 2011. http://www.sciencemag.org/content/278/5344/1734.full. doi: 10.1126/science.278.5344 .1734.

Soffen, Gerald A. "Scientific Results of the Viking Missions." *Science* 194, no. 4271 (December 1976): 1274–76. Accessed 2011. http://www.sciencemag.org/content/194/4271/1274.abstract. doi: 10.1126/science.194.4271.1274.

Space News. "Mars Science Lab Needs $44m More to Fly, NASA Audit Finds." June 8, 2011. Accessed 2011. http://www.spacenews.com/civil/110608-msl-needs-more-nasa-audit.html.

University of Arizona. "Phoenix Mars Lander." Accessed 2011. http://phoenix.lpl.arizona.edu/index.php.

GENERAL REFERENCES

Caltech CODA: Oral Histories Online (used by permission): http://oralhistories.library.caltech.edu/.

European Space Agency: http://www.esa.int/.

Mars Society (for general information on FMARS and Earth-based Mars research): http://www.marssociety.org/.

National Aeronautics and Space Administration (NASA)/ *Astrobiology Magazine*: http://www.astrobio.net/.

National Aeronautics and Space Administration (NASA)/Jet Propulsion Laboratory (JPL) BEACON Information Commons: http://beacon.jpl.nasa.gov/site-index.

National Aeronautics and Space Administration (NASA)/ Jet Propulsion Laboratory (JPL) Mars Exploration Program: http://mars.jpl.nasa.gov/.

National Aeronautics and Space Administration (NASA)/Johnson Space Center, History Portal: http://www.jsc.nasa.gov/history/oral_histories/participants.htm.

National Aeronautics and Space Administration (NASA)/Lunar and Planetary Institute: http://www.lpi.usra.edu/.

National Aeronautics and Space Administration (NASA)/National Space Science Data Center: http://nssdc.gsfc.nasa.gov/.

National Archives: http://www.archives.gov/.

National Science Foundation (for Antarctic research and projects): http://www.nsf.gov/.

Niels Bohr Library and Archives, American Institute of Physics (used by permission): http://www.aip.org/history/ohilist/.

INDEX

Illustrations and photos are indicated by **bold** page numbers.